A DICTIONARY OF MOSSES

an alphabetical listing of genera
indicating familial disposition, nomenclatural
and taxonomic synonymy
together with
a systematic arrangement of the
families of mosses
and
a catalogue of family names
used for mosses

Second Printing, with corrections and additions

MARSHALL R. CROSBY
ROBERT E. MAGILL

SCIENTIFIC
PUBLISHERS

Published by:

SCIENTIFIC PUBLISHERS (INDIA)
5A, New Pali Road, P.O. Box 91
Jodhpur 342 001, India
E-mail: info@scientificpub.com
Website: www.scientificpubonline.com

Print 2026

First published in India 1987

ISBN: 978-81-85046-65-5
eISBN: 978-93-88449-58-8

Printed in India

CONTENTS

Introduction

Although there are probably more species of mosses than any other class of green plants except angiosperms, their classification above the level of genus has received surprisingly little attention during the past SO years. However, during that period a large number of genera have been described as new, some generic nameshave been reduced to synonymy of others, while still others have been replaced on nomenclatural grounds. A few genera have been reclassified from one family to another in attempts to arrive at a better overall classification of the mosses. This DICTIONARY provides a means of quickly determining the family in which any genus of mosses is placed. Since names treated as synonyms, many orthographic variants, and the names for genera which are accepted as taxonomic entities are included, the DICTIONARY should prove a useful curatorial tool. The last such comprehensive guide to the familial disposition to the genera of mosses appeared more than SO years ago as Brotherus' treatment of the mosses in *Die natürlichen Pflanzenfamilien* (1924-25). The DICTIONARY began as a simple card catalogue of the generic names in Brotherus and was used as a curatorial too! at the Missouri Botanical Garden. But this basic list soon proved inadequate, since so many moss generic names have been published since 1925. Using *IndexMuscorum* (Wijk et al., 1959-69) as a source, the original publications of all post-1925 names for moss genera were studied to determine the family in which the genera were originally placed. Later, all the remaining validly published names listed in the *Index Muscorum* were added to the catalogue, these being basically synonyms and orthographic variants. Regular surveys of the current literature supplied many additional generic names which have been published since the closing date for the *Index Muscorum,* the beginning of 1963. Finally, the catalogue was compared against a printout of the moss names which have been catalogued for the *Index Nominum Genericorum* project.

In the DICTIONARY OF MOSSES proper the familial disposition used by Brotherus is always given. But in those cases where we think that it has been reasonably demonstrated that a genus should be placed in another family, we have done so, indicating Brotherus' placement in parentheses. For genera published since 1925, the disposition of the original publishing author is given, unless it has been changed subsequently. In these cases, the original disposition is given in parentheses.

Only names which have been validly published are included in the DICTIONARY. Generic names which are considered correct appear in *italic* while those which are not appear in roman. Taxonomic synonymy is indicated by the equal to sign (=), and nomenclatural synonymy is indicated by the identical with sign (≡). Many generic names have never been typified, and in some of these cases it is not yet possible to give a correct name. The notation "= p.p. ... & p:p. . . ." has been used when the species of untypified names arc. now placed in two or more genera. Later homonyms are indicated by the abbreviation "hom. illeg." Names which are otherwise illegitimate are noted by "nom. illeg." We have generally not distinguished between intentional and unintentional changes in orthography. These changes have been treated as orthographic variants ("orthogr. var.") and attributed to the author who made the change. In those few cases where the original orthography of a name has been changed and the changed orthography is the accepted one, the changed orthography is attributed to the original publishing author of the name rather than the person who made the correction. In these cases the original spelling is listed as a variant. Names conserved or rejected under the *International Code of Botanical Nomenclature* are indicated with "nom. cons." and "nom. rejic.." respectively. For a few genera which are recognized as good taxonomic groups we have been unable to determine correct names.

Several revised classifications of mosses at the family level have appeared since 1925. The revisions have been slight, and most floras and many herbaria are arranged according to the system used by Brotherus. We have included an outline of this system, the SYSTEMATIC ARRANGEMENT OF THE FAMILIES OF MOSSES, as an aid to curating herbaria arranged according to this system. This SYSTEMATIC ARRANGEMENT is modified from that used by Brotherus to the extent of the addition of families which have been published since 1925 and by the addition of some pre-1925 families which we recognize as distinct but which he did not. The families are numbered sequentially according to the system used by Brotherus; names which have been intercalated into this system are given the number of the closest *fireceding* family in Brotherus, followed by a letter as needed. In the SYSTEMATIC ARRANGEMENT, families which are recognized in the DICTIONARY are in small caps (DICRANACEAE), while those which are not recognized are in lowercase (Leucobryaceae).

An alphabetical INDEX TO THE SYSTEMATIC ARRANGEMENT is provided so that the various families can be easily located in the SYSTEMATIC ARRANGEMENT. The number after each entry in the INDEX is that given to the family in the SYSTEMATIC ARRANGEMENT.

A second alphabetical listing of family names is also included. This CATALOGUE OF FAMILY NAMES USED FOR MOSSES gives the full citation for all of the validly published names for families of mosses. Those names which are adopted in our SYSTEMATIC ARRANGEMENT, and hence in the DICTIONARY, appear in small caps, while those which are not used are in lower case. The correct name for these latter names is indicated with the equal to or identical with sign. A few names which we consider invalid are also included in this CATALOGUE, for reasons explained below.

Three criteria, based on the Principles and Articles of the *Code* have been used to determine if a name is included in the CATALOGUE. First, the rank given to the name must have been specifically stated as family or order, or variations of these words (Article 18). Second, we have interpreted Principle V and Article 32, Note 1, to mean that a name must have had a Latin ending when published in order to be validly published. Compilers of catalogues of family names for some other groups of plants have interpreted the *Code* more broadly than we and accepted as validly published names with modem endings. The modem endings were changed to the correct -aceae Latin termination by these other compilers. Third, the name must have been validly published under Articles 18 or 32. Several names which are not based on generic names are included in the CATALOGUE. We consider these invalid under Article 18, Note 3, but have included them since they have been widely used in the literature.

Most of the family names in current usage are correct under the *Code*. There are a few important exceptions, all based on the principle of priority: Cerato- dontaceae and Distichiaceae were published simultaneously (1860) and antedate Ditrichaceae (1887); Hyophilaceae (1847) antedates Pottiaceae (1856); Macro- hymeniaceae (1883) antedates Sematophyllaceae (1908); and Orthotheciaceae (1856) and Pterigynandraceae (1876) antedate Entodontaceae (1897). One of us has proposed that the next International Botanical Congress conserve the more familiar names and reject their earlier synonyms (Magill, 1977, 1978).

Authors are abbreviated according to Sayre, Bonner, and Culberson (1964) throughout. Abbreviations for authors not included in their list have been derived in accordance with the principles used there.

Acknowledgments. This work was carried out during the tenure of National Science Foundation grants GB-28497 and GB-43656 and National Endowment for the Arts grant R50-20-295, all to the Missouri Botanical Garden. We, of course, accept responsibility for the final product, but copies of the manuscript were sent to several reviewers, and we appreciate the many useful suggestions made by H. A. Crum, W. C. Steere, R. H. Zander, A, Touw, H. Robinson, A. Noguchi, W. Schultze-Motel, W. A. Weber, W. B. Schofield, Z. Iwatsuki, G. L. Smith, G. Sayre, M. G. Manuel, N. G. Miller, T. Koponen, and W. D. Margadant. We thank Ellen Farr, Smithsonian Institution, for making a printout of the names for the *Index Nominum Genericarum* available to us and for useful suggestions. Martha Botsford Ballard and Susan Stangler, Missouri Botanical Garden, pointed out several omissions and helped proofread the manuscript. Judy Wolf saw the main portion of the manuscript through several retypings.

We particularly appreciate the encouragement received from Donald Richards and dedicate the DICTIONARY to him.

Acknowledgements for the second printing: The second printing lists several new names which have been published since the first printing and reflects some changes in taxonomic concepts which have appeared in the literature. We recognize 795 genera as distinct; 792 were recognized in the original printing. Several errors have been corrected, and we appreciate the help of the following for bringing some of these to our attention: M. Botsford Ballard, W. R. Buck, Z. Iwatsuki, T. Koponen, and R. H. Zander.

DICTIONARY OF MOSSES

Abietinella C. Müll .... THUIDIACKAE
Acamptodous Duby = *Lepidopilum* .... DALTONIACEAE
Acanthocladiella Fleisch .... SEMATOPHYLLACEAE
Acanthocladium Mitt., hom. illeg. ≡ *Wijkia* .... SEMATOPHYLLACEAE
Acanthodium Mitt., hom. illeg. = *Trismegistia* .... SEMATOPHYLLACEAE
Acantholoma Herz., hom. illeg. ≡ *Bryoacanthaloma* .... RHIZOQONIACEAE
Acanlhorrhynchium Fleisch .... SEMATOPHYLLACEAE
Acaulon C. Müll .... POTTIACEAE
Achrohypnella Herz. = *Sauloma* .... HOOKERIACEAE
Achrolepis (Lindb.) Card., nom. illeg. ≡ *Myuroclada* .... BRACHYTHECIACEAE
Achrophyllum Vitt & Crosby .... HOOKERIACEAE
Achyrophyllum Shaw ≡ *Ptychomnion* .... PTYCHOMNUCEAE
Acidodontium Schwaegr., nom. cons. .... BRYACEAE
Acosta C. Müll., hom. illeg. = *Vesicularia* .... HYPNACEAE
Acritodon Robins. .... SEMATOPHYLLACEAE
Acrodadiopsis (Broth.) Card. = *Catagonium* .... PHYLLOGONIACEAE
Acrocladium Mitt .... AMBLYSTEGIACEAE (LEMBOPHYLLACEAE)
Acrocryphaea B.S.G. ex Broth., nom. illeg. ≡ *Schoenobryum* .... CRYPHAEACEAE
Acroporium Mitt. = *Macrohymenium* .... SEMATOPHYLLACEAE
Acroschisma (Hook. f. & Wils.) Lindl. = *Andreaea* .... ANDREAEACEAE
Actinodontium Schwaegr. .... DALTONIACEAE (Hookeriaceae)
Actinostoma Reim. & Sak. = *Entodon* .... ENTODONTACEAE
Actinothuidium (Besch.) Broth. .... THUIDIACEAE
Adelothecium Mitt. .... HOOKERIACEAE
Aërobryidium Fleisch. in Broth. .... METEORIACEAE
Aerobryopsis Fleisch .... METEORIACEAE
Aërobryum Dozy & Molk. .... METEORIACEAE
Aligrimmia Williams .... GRIMMIACEAE
Allioniella Broth .... SEMATOPHYLLACEAE
Allorgea Ando .... HYPNACEAE
Aloïdella Vent., nom. rejic. = *Alaïna* .... POTTIACEAE
Aloïna Kindb., nom. cons. .... POTTIACEAE
Aloïnella Card. .... POTTIACEAE
Alophosia Card. .... POLYTRICHACEAE
Alsia Sull. .... CRYPHAEACEAE
Amblyodon B.S.G., nom. cons. .... MEESIACEAE
Amblyodum P. Beauv., nom. rejic. ≡ *Amblyodon* .... MEESIACEAE
Amblyphyllum Lindb., nom. illeg. ≡ *Splachnobryum* .... POTTIACEAE
Amblystegiella Loeske ≡ *Plaiydictya* .... AMBLYSTEGIACEAE
Amblystegium B.S.G. .... AMBLYSTEGIACEAE
Amblytropis (Mitt.) Broth. .... DALTONIACEAE (Hookeriaceae)
Amphidium Schimp., nom. cons. .... ORTHOTRICHACEAE (Dicranaceac)
Amphidium Nees in Sturm., nom. rejic. = *Zygodon* .... ORTHOTRICHACEAE
Amphidrium Steud., orthogr. var. Amphidium Nees = *Zygodon* .... ORTHOTRICHACEAE
Amphoridium Schimp., hom. illeg. ≡ *Amphidium* .... ORTHOTRICHACEAE
Amphoritheca Hampe = *Entosthodon* .... FUNARIACEAE
Anacalypta Roehl. ex Fürnr. = *Pottia* .... POTTIACEAE

Genus	Family
Anacamptodon Brid.	FABRONIACEAE
Anacolia Schimp., nom. cons.	BARTRAMIACEAE
Ancistrodes Hampe	METEORIACEAE
Andreaea Hedw.	ANDREAEACEAE
Andreaeobryum Steere & B. Murr.	ANDREAEOBRYACEAE
Angstroemia B.S.G., orthogr. var. *Aongstroemia*	DICRANACEAE
Anhymenium Griff. = *Regmatodon*	REGMATODONTACEAE
Anictangium Hedw., nom. rejic. = *Anoectangium*	POTTIACEAE
Anisodon B.S.G. = *Clasmatodon*	FABRONIACEAE
Anisostichium Mitt. = *Epipterygium*	BRYACEAE
Anisothecium Mitt.	DICRANACEAE
Anodon Rabenh., hom. illeg. = *Schistidium*	GRIMMIACEAE
Anodontium Brid., nom. rejic. = *Drummondia*	ORTHOTRICHACEAE
Anodus B.S.G. = *Seligeria*	SELIGERIACEAE
Anoectangium Schwaegr., nom, cons.	POTTIACEAE
Anoectangium Lindb., hom. illeg. = *Amphidium*	ORTHOTRICHACEAE
Anomobryopsis Card.	BRYACEAE
Anomobryum Schimp.	BRYACEAE
Anomodon Hook. & Tayl.	THUIDIACEAE
Antitrichia Brid.	LEUCODONTACEAE
Anyctangium Lam. & Cand., orthogr. var. Anictangium = *Anoectangium*	POTTIACEAE
Aongstroemia B.S.G.	DICRANACEAE
Aongstroemiopsis Fleisch.	DICRANACEAE
Apalodium Mitt. = *Orthodontium*	BRYACEAE
Aphanorhegma Sull. in Gray	FUNARIACEAE
Apiocarpa Hüb. = *Mielichhoferia*	BRYACEAE
Aplodon R. Br.	SPLACHNACEAE
Apocarpium Desv. = *Grimmia*	GRIMMIACEAE
Apocarpum P. Beauv. = *Grimmia*	GRIMMIACEAE
Apodanthus B. Pyl. = *Splachnum*	SPLACHNACEAE
Aporella Podp. = *Micromitrium*	EPHEMERACEAE
Aptchella Bartr., orthogr. var. *Aptychella*	SEMATOPHYLLACEAE
Apterygium Kindb. = *Platydictya*	AMBLYSTEGIACEAE
Aptychella (Broth.) Herz.	SEMATOPHYLLACEAE
Aptychopsis (Broth.) Fleisch.	SEMATOPHYLLACEAE
Aptychus (C. Mull.) C. Müll. = *Sematophyllum*	SEMATOPHYL LACEAE
Arbuscula Crum, Steere & Anderson, nom. illeg. =*Thamnobryum*	NECKERACEAE
Archboldiella Bartr. = *Distichophyllidium*	DALTONIACEAE
Archephemeropsis Renn. = *Ephemeropsis*	HOOKERIACEAE
Archidiella Irmsch. = *Archidium*	ARCHIDIACEAE
Archidium Brid.	ARCHIDIACEAE
Arctoa B.S.G.	DICRANACEAE
Argyrobryum (C. Müll.) Kindb. = *Bryum*	BRYACEAE
Arrhenopterum Hedw., nom. rejic. = *Aulacomnium*	AULACOMNIACEAE
Arthrocormus Dozy & Molk.	CALYMPERACEAE (Leucobryaceae)
Aschisma Lindb.	POTTIACEAE
Aschistodon Mont., nom. rejic. = *Ditrichum*	DITRICHACEAE
Asteriscium (C. Mull.) Hilp., hom. illeg. = *Trichostomopsis*	POTTIACEAE
Asterodontium Dix. & Gepp, orthogr. var. Astrodontium Schwaegr. = *Leucodon*	LEUCODONTACEAE

Asiomiopsis C. Müll.	DITRICHACEAE
Astomum Hampe	POTTIACEAE
Astrodontium Dozy & Molk., hom. illeg. ≡ *Clastobryum*	SEMATOPHYLLACEAE
Astrodontium Schwaegr. = *Leucodon*	LEUCODONTACEAE
Astrophyllum Lindb , nom. illeg. ≡ *Mnium*	MNIACEAE
Atractylocarpus Mitt., nom. cons.	DICRANACEAE
Atrichopsis Card.	POLYTRICHACEAE
Atrichum P. Beauv., nom. cons.	POLYTRICHACEAE
Aulacomitrium Mitt. = *Glyphomitrium*	PTYCHOMITRIACEAE
Aulacomitrium Broth, ex C. Müll., hom. illeg. = *Glyphomitrium*	PTYCHOMITRLACEAE
Aulacomnion Schwaegr. ≡ *Aulacomnium*	AULACOMNIACEAE
Aulacomnium Schwaegr., nom. cons.	AULACOMNIACEAE
Aulacopilum Wils.	ERPODIACEAE
Austinella Williams = *Oncophorus*	DICRANACEAE
Austinia C.Müll.	FABRONIACEAE

Balwiniella Broth. ex Fleisch.	NECKERACEAE
Barbella Fleisch. in Broth	METEORIACEAE
Barbellopsis Broth	METEORIACEAE
Barbiferus Poir., nom. illeg. ≡ *Barbula*	POTTIACEAE
Barbula Hedw., nom. cons.	POTTIACEAE
Bamesia Card.	POTTIACEAE
Barthramia Gaertn., Meyer & Schreb., orthogr. var. *Bartramia*	BARTRAMIACEAE
Bartleya Robins. = *Dicranella*	DICRANACEAE
Bartramia Hedw., nom. cons.	BARTRAMIACEAE
Bartramidula B.S.G., nom. cons.	BARTRAMIACEAE
Bartramiopsis Kindb.	POLYTRICHACEAE
Beccaria C. Müll. = *Pottia*	POTTIACEAE
Beckettia C. Müll. = *Pottia*	POTTIACEAE
Beddomiella Dix. (Pottiaceae) = *Thiemia*	DICRANACEAE
Bellia Broth., hom. illeg. ≡ *Crosbya*	HOOKERIACEAE
Bellibarbula Chen	POTTIACEAE
Bergia Fürnr., hom. illeg. = *Entosthodon*	FUNARIACEAE
Bescherellia Duby	CYRTOPODACEAE
Bestia Broth.	CRYPHAJEACEAE
Bissetia Broth. in Fleisch.	NECKERACEAE
Bizotia Pierr. = *Campylopus*	DICRANACEAE
Blindia B.S.G.	SELIGERIACEAE
Blindiopsis Bartr. & Dix. = *Blindia*	SELIGERIACEAE
Boulaya Card.	THUIDIACEAE
Brachelyma Schimp. ex Card.	FONTINALACEAE
Brachydontium Fürnr.	SELIGERIACEAE
Brachyelyma Hag., orthogr. var. *Brachelyma*	FONTINALACEAE
Brachymeniopsis Broth.	FUNARIACEAE
Brachymenium Schwaegr.	BRYACEAE
Brachymitrion Tayl.	SPLACHNACEAE
Brachyodon Fürnr., orthogr. var. Brachyodus = *Brachydontium*	SELIGERIACEAE
Brachyodontium Broth., orthogr. var. *Brachydontium*	SELIGERIACEAE

Brachyodus Fümr. ex Nees & Homsch., nom. illeg. = *Brachydontium* .... SELIGERIACEAE
Brachypodium Brid., bom. illeg. = *Ptychomitrium* .... PTYCHOMITRIACEAE
Brachysteleum Reichenb., nom. rejic,. = *Ptychomitrium* .... PTYCHOMITRIACEAE
Brachythecium B.S.G. .... BRACHYTHECIACEAE
Brachytrichum Röhl. = *Orthotrichum* .... ORTHOTRICHACEAE
Braithwaitea Lindb. .... HYPNODENDRACEAE
Braunfelsia Par. .... DICRANACEAE
Braunia B.S.G. .... HEDWIGIACEAE
Braunia Hampe, hom, illeg. ≡ *Dendrocryphaea* .... CRYPHAEACEAE
Breidleria Loeske = *Hypnum* .... HYPNACEAE
Breutelia (B.S.G.) Schimp. .... BARTRAM IACEAE
Brium Brid., orthogr. var. *Bryum* .... BRYACEAE
Brothera C. Müll. .... DICRANACEAE
Brotherella Loeske ex Fleisch .... SEMATOPHYLLACEAE
Brotherobryum Fleisch. .... DICRANACEAE
Bruchia Schwaegr. .... DICRANACEAE
Bruchia Homsch., bom. illeg.
Bryella Berk. = *Pottia* .... POTTIACEAE
Bryhnia Kaur. .... BRACHYTHECIACEAE
Bryoacantholoma Schultze-Mot. .... RHIZOGONIACEAE
Bryoandersonia Robins. .... BRACHYTHECIACEAE
Bryobartramia Sainsb. .... BRYOBARTRAMIACEAI
Bryobrittonia Williams .... ENCALYPTACEAE (Pottiaceae)
Bryobrothera Thér. .... RHIZOGONIACEAE
Bryoceuthospora Crum & Anderson .... POTTIACEAE
Bryodixonia Sainsb. .... ORTHOTRICHACEAE
Bryodusenia Robins. = *Ancistrodes* .... METEORIACKAE
Bryoerythrophyllum Chen .... POTTIACEAE
Bryohumbertia P. de la Varde & Thér. .... DICRANACEAE
Bryomanginia Thér. .... DITRICHACEAE
Bryomnium Card. in Broth. (Mniaceae) = *Tayloria* .... SPLACHNACEAE
Bryomyces Miq. = *Orthotrichum* .... ORTHOTRICHACEAE
Bryonoguchia Iwats. & H. Inoue .... THUIDIACEAE
Bryoporteria Ther. = *Camptodontium* .... DICRANACEAE
Bryosedgwickia Card. & Dix. .... HYPNACEAE
Bryotestua Ther. & P. de la Varde .... DICRANACEAE
Bryowijkia Nog. .... HEDWIGIACEAE
Bryoxiphium Mitt., nom. cons. .... BRYOXIPHIACEAE
Bryoziphium Mitt., orthogr. var. *Bryoxiphium* .... BRYOXIPHIACEAE
Bryum Hedw. .... BRYACEAE
Bucklandia Roiv., hom. illeg. = *Racomitrium* .... GRIMMIACEAE
Bucklandiella Roiv. = *Racomitrium* .... GRIMMIACEAE
Burnettia Grout, hom. illeg. ≡ *Homalotheciel* .... *la* BRACHYTHECIACEAE
Buxbaumia Hedw. .... BUXBAUMIACEAE

Callibryon Zenk. & Dietr., nom. illeg. ≡ *Atrichum* .... POLYTRICHACEAE

Callicladium Crum .... HYPNACEAE
Callicosta C. Müll. .... DALTONIAGEAE
Callicostella (C. Müll.) Mitt. = *Schizomitrium* .... HOOKERIACEAE

Genus	Family
Callicostellopsis Broth. (Hookeriaceae) = *Helicoblepharum*	DALTONIACEAE
Calliergidium Grout	AMBLYSTEGIACEAE
Calliergon (Sull.) Kindb.	AMBLYSTEGIACEAE
Calliergonella Loeske	AMBLYSTEGIACEAE
Calliergonella (Card.) Card. & Broth., hom. illeg. = *Catagonium*	PHYLLOGONIACEAE
Calohypnum Sak. = *p.p.Brotherella* (Sematophyllaceae) & p.p. *Hypnum* (Hypnaceae)	
Calomnion Hook. f. & Wils	CALOMNIACEAE
Calomnium Broth., orthogr. var. *Calomnion*	CALOMNIACIAE
Calymperes Sw. in Web	CALYMPERACEAE
Calymperidium Dozy & Molk. = *Syrrhopodon*	CALYMPERACEAE
Calymperiopsis R. S. Chopra, orthogr. var. *Calymperopsis*	CALYMPERACEAE
Calymperopsis (C. Müll.) Fleisch	CALYMPERACEAE
Calyptopogon (Mitt.) Broth	POTTIACEAE
Calyptothecium Mitt	PTEROBRYACEAE (Neckeraceae)
Calyptrochaeta Desv	HOOKERIACEAE
Calyptrochaete Wijk, Marg. & Florsch., orthogr. var. *Calyptrochaeta*	HOOKERIACEAE
Campilopus Henr., orthogr. var. *Campylopus*	DICRANACEAE
Camplothecium Schur, orthogr. var. Camptothecium =*Homalothecium*	BRACHYTHECIACEAE
Campothecium Kindb., orthogr. var. Camptothecium ≡ *Homalothecium*	BRACHYTHECIACEAE
Camptochaete Reichardt	LEMBOPHYLLACEAE
Camptodontium Dus	DICRANACEAE
Camptothecium B.S.G. ≡ *Homalothecium*	BRACHYTHECIACEAE
Campyliadelphus (Kindb.) R. S. Chopra	AMBLYSTEGIACEAE
Campylium (Sull.) Mitt	AMBLYSTEGIACEAE
Campylocarpus Grout, orthogr. var. *Campylopus*	DICRANACEAE
Campylochaetium Besch. = *Dicranella*	DICRANACEAE
Campylodontium Schwaegr. = *Anacamptodon*	FABRONIACEAE
Campylodontium Dozy & Molk., hom. illeg. = *Mesonodon*	ENTODONTACEAE
Campylophyllum (Schimp.) Fleisch	AMBLYSTEGIACEAE
Campylopodiella Card. = *Brothera*	DICRANACEAE
Campylopodium(C. Müll.) Besch	DICRANACEAE
Campylopus Brid	DICRANACEAE
Campylopus Wahlenb., hom. illeg. = *Grimmia*	GRIMMIACEAE
Campylostegium Hampe, orthogr. var. *Campylostelium*	PTYCHOMITRIACEAE
Campylosteleum Solms, orthogr. var. *Campylostelium* .	PTYCHOMITRIACEAE
Campylostelium B.S.G	PTYCHOMITRIACEAE
Campylostylium Mont., orthogr. var. *Campylostelium* .	PTYCHOMITRIACEAE
Campylotrichum Sikora	?HOOKERIACEAE
Cardotia Besch. in Card. (Leucobryaceae) = *Leucobryum*	DICRANACEAE
Carinqfolium Williams (Leucobryaceae)	CALYMPERACEAE
Carovaglia Endl., orthogr. var. *Garovaglia*	PTEROBRYACEAE
Carpoecia Vent. = *Campylopus*	DICRANACEAE
Catagoniopsis Broth	BRACHYTHECIACEAE

Catagonium C. Müll, ex Broth	PHYLLOGONIACEAE
Catagonium Broth., hom. illeg. (Plagiotheciaceae) = *Catagonium*	PHYLLOGONIACEAE
Catascopium Endl., orthogr. var. *Catoscopium*	CATOSCOPIACEAE
Catharinea Web. & Mohr, nom. rejic. = *Atrichum*	POLYTRICHACEAE
Catharinella (C. Mull.) Kindb. = *Pogonatum*	POLYTRICHACEAE
Catharomnion Hook. f. & Wils	DALTONIACEAE (Hypopterygiaceae)
Catharomnium Broth., orthogr. var. *Catharomnion*	DALTONIACEAE
Catogoniopsis Broth., orthogr. var. *Catagoniopsis*	BRACHYTHECIACEAE
Catoptridium Brid. = *Schistostega*	SCHISTOSTEGACEAE
Catoscopium Brid	CATOSCOPIACEAE
Cavanillea Borkh., hom. illeg. = *Weissia*	POTTIACEAE
Cecalyphum P. Beauv., nom. illeg. ≡ *Dicrarum*	DICRANACEAE
Cephalocladium Laz	SEMATOPHYLLACEAE
Cephalotrichum B.S.G., nom. illeg. = *Pogonatum*	POLYTRICHACEAE
Cepbaloxis P. Beauv. ≡ *Bartramia*	BARTRAMIACEAE
Ceratodon Brid	DITRICHACEAE
Ceuthospora Crum & Anderson, hom. illeg. ≡ *Bryoceuthospora*	POTTIACEAE
Chaetephora Brid., hom. illeg. ≡ *Calyptrochaeta*	HOOKERIACEAE
Chaetomitrella Hampe = *Chaetomitrium*	HOOKERIACEAE
Chaetomitriopsis Fleisch	HOOKERIACEAE
Chaetomitrium Dozy & Molk	HOOKERIACEAE
Chaetophora Brid., hom. illeg. ≡ *Calyptrochaeta*	HOOKERIACEAE
Chamaebryum Thén & Dix	GIGASPERMACEAE
Chamberlainia Grout	BRACHYTHECIACEAE
Cheilothela Lindb	DITRICHACEAE
Chilothela Hag., orthogr. var. *Cheilothela*	DITRICHACEAE
Chionobryum Glow. = *Brachythecium*	BRACHYTHECIACEAE
Chionoloma Dix	POTTIACEAE
Chinostomum C. Müll	SEMATOPHYLLACEAE
Chloronotus Vent., nom. rejic. = *Crossidium*	POTTIACEAE
Chorisodontium (Mitt.) Broth	DICRANACEAE
Chrystoblastella Williams (Pottiaceae) = *Cheilothela*	DITRICHACEAE
Chrysocladium Fleisch	METEORIACEAE
Chryso-hypnum Hampe = *Mittenothamnium*	HYPNACEAE
Chrysohypnum G. Roth, hom. illeg. = *Campylium*	AMBLYSTEGIACEAE
Chrysypnum Hag., orthogr. var. Chryso-hypnum Hampe = *Mittenothamnium*	HYPNACEAE
Cicclidotus P. Beauv., orthogr. var. *Cinclidotus*	POTTIACEAE
Cicelidotus Clairv., orthogr. var. *Cinclidotus*	POTTIACEAE
Cinclidium Sw. in Schrad	MNIACEAE
Cinclidotus P. Beauv	POTTIACEAE
Cirriphyllym Grout	BRACHYTHECIACEAE
Cladastomum C. Mull	DITRICHACEAE
Cladodium Brid. = *Bryum*	BRYACEAE
Cladomnion Hook. f. & Wils	PTYCHOMNIACEAE
Cladomniopsis Fleisch	PTYCHOMNIACEAE
Cladomnium Hook. f. & Wils., orthogr. var. *Cladomnion*	PTYCHOMNIACEAE
Cladophascum Dix. in Sim	POTTIACEAE (Dicranaceae)

Cladopodanthus Dozy & Molk .... DICRANACEAE (Leucobryaceae)
Cladostomum C. Müll., orthogr. var. *Cladostomum*.................. DITRICHACEAE
Claopodium (Lesq. & James) Ren. & Card .... THUIDIACEAE
Clasmatodon Hook. & Wils .... FABRONIACEAE
Clastobryella Fleisch.. SEMATOPHYLLACEAE
Clastobryophilum Fleisch.. SEMATOPHYLLACEAE
Clastobryopsis Fleisch. = *Aptychella* .,..................................... SEMATOPHYLLACEAE
Clastobryum Dozy & Molk .... SEMATOPHYLLACEAE
Cleistostoma Brid., nom. illeg. ≡ *Syrrhopodon* .... CALYMPERACEAE
Cleistostoma sensu Broth. ≡ *Bryowijkia*.................................. HEDWIGIACEAE
Climacium Web. & Mohr .... CLIMACIACEAE
Climatium Gümb., orthogr. var. *Climacium* .... CLIMACIACEAE
Clistostoma Hag., orthogr. var. Cleistostoma ≡ *Syrrhopodon* .... CALYMPERACEAE
Cnestrum I. Hag .... DICRANACEAE
Codonoblepharon Schwaegr. = *Zygodon* .... ORTHOTRICHACEAE
Codonoblepharum Brid., orthogr. var. Codo noblepharon = *Zygodon*.. ORTHOTRICHACEAE
Codonoblepharum Dozy & Molk., hom. illeg. = *Mitthyridium*... CALYMPERACEAE
Codonophorus Desv. = *Racomitrium* .... GRIMMIACEAE
Codriophorus P. Beauv. = *Racomitrium* .... GRIMMIACEAE
Coelidium (Hook. f. & Wils.) Reichardt, hom. illeg. *Weymouthia* .... METEORIACEAE
Coleochaetium (Besch.) Ren. & Card., nom. illeg. = *Leiomitrium* .... ORTHOTRICHACEAE
Collania Sak., orthogr. var. *Gollania*.. RHYTIDIACEAE
Colobodontium Herz .... SEMATOPHYLLACEAE
Complanato-hypnum Hampe = *Stereophyllum* .... PLAGIOTHECIAPEAE
Conordia Robins .... AMBLYSTEGIACEAE
Conomitrium Mont., nom. illeg. = *Fissidens* .... FISSIDENTACEAE
Conostomum Sw. in Web. & Mohr.. BAKTRAMIACEAE
Coscinodon Spreng.. GRIMMIACEAE
Coscinodontella Williams .... GRIMMIACEAE
Costesia Ther.. FUNARIACEAE
Craspedophyllum (Card.) Grout, hom. illeg. *Groutiella* .... ORTHOTRICHACEAE
Crastobryella Horik. & Ando, orthogr. var. *Clastobryella* .... SEMATOPHYLLACEAE
Cratoneurella Robins .... BRACHYTHECIACEAE
Cratoneuron (Sull.) Spruce .... AMBLYSTEGIACEAE
Cratoneuropsis (Broth.) Fleisch .... AMBLYSTEGIACEAE
Cratoneurum Hag., orthogr. var. *Cratoneuron* .... AMBLYSTEGIACEAE
Crepidophyllum Herz .... HYPNACEAE
Cribrodontium Herz .... ENTODONTACEAE
Crosbya Vitt .... HOOKERIACEAE
Crossidium Jur., nom. cons .... POTTIACEAE
Crossomitrium C. Müll... DALTONIACEAE (Hookeriaceae)
Crumia Schof .... POTTIACEAE
Cryphaea Mohr in Web .... CRYPHAEACEAE
Cryphaeadelpbus (C. Müll.) Card., nom. illeg. = *Brachelyma* .... FONTINALACEAE

Cryphaeophilum Fleisch	METEORIACEAE
Cryphidium (Mitt.) Jaeg. (Neckeraceae) = *Cryphaea*	CRYPHAEACEAE
Cryptangium C. Müll., hom. illeg. ≡ *Hydropogonella*	HYDROPOGONACEAE
Cryptocarpella Trev., nom. illeg. ≡ *Desmotheca*	ORTHOTRICHACEAE
Cryptocarpon Dozy & Molk., orthogr. var. Cryptocarpus ≡ *Desmotheca*	ORTHOTRI CHACEAE
Cryptocarpus Dozy & Molk., hom. illeg. ≡ *Desmotheca*	ORTHOTRICHACEAE
Cryptodicranum Bartr	DICRANACEAE
Cryptogonium (C. Mull.) C. Müll, in F. Müll. ≡ *Phyllogonium*	PHYLLOGONIACEAE
Cryptoleptodon Ren. & Card	NECKERACEAE
Cryptoneurum Thér & P. de la Varde	RHYTIDIACEAE
Cryptopodia Röhl. = *Neckera*	NECKERACEAE
Cryptopodium Brid	RHIZOGONIACEAE
Cryptopodium (Röhl.) Fürnr, hom. illeg. = *Neckera*	NECKERACEAE
Cryptoseta (Amott) Hook, ex Kitt. ≡ *Cryptopodium*	RHIZOGONIACEAE
Cryptotheca Hornsch. in Mitt., hom. illeg. = *Symphysodon*	PTEROBRYACEAE
Ctenidiadelphus Fleisch	HYPNACEAE
Ctenidium (Schimp.) Mitt	HYPNACEAE
Ctenium C. Jens., nom. illeg. ≡ *Ptilium*	HYPNACEAE
Cupressina C. Müll., nom. illeg. ≡ *Hypnum*	HYPNACEAE
Cuspidaria C. Müll., hom. illeg. = *Entodon*	ENTODONTACEAE
Cyathophorella (Broth.) Fleisch	DALTONIACEAE (Hypopterygiaceae)
Cyathophorum P. Beauv	HOOKERIACEAE (Hypopterygiaceae)
Cyathothecium Dix	HYLOCOMIACEAE
Cyclodictyon Mitt	HOOKERIACEAE
Cycnea Berk., hom. illeg. = *Phascum*	POTTIACEAE
Cylicocarpus Lindb. = *Amphidium*	ORTHOTRIC HACEAE
Cylindrothecium B.S.G., nom. illeg. = *Entodon*	ENTODONTACEAE
Cymbaria Tayl., hom. illeg. = *Eustichia*	EUSTICHIACEAE
Cymbifoliella Dix. = *Entodon*	ENTODONTACEAE
Cynodon Brid., hom. illeg. = *Distichium*	DITRICHACEAE
Cynodontiella (Limpr.) Bryhn = *Cynodontium*	DICRANACEAE
Cynodontium Schimp., nom. cons	DICRANACEAE
Cynodontium Brid., nom. rejic. = *Distichium*	DITRICHACEAE
Cynodontium B.S.G., nom. rejic. = *Oreoweisia*	DICRANACEAE
Cynontodium Hedw., nom. rejic. = *Distichium*	DITRICHACEAE
Cyptodon (Broth.) Par. & Schimp. ex Fleisch	CRYPHAEACEAE
Cyptodontopsis Dix	CRYPHAEACEAE
Cyrtodon (R. Br.) Hook. = *Tayloria*	SPLACHNACEAE
Cyrto-hypnum Hampe = *Thuidium*	THUIDIACEAE
Cyrtomnium Holmen	MNIACEAE
Cyrtopodendron Fleisch	CYRTOPODACEAE
Cyrtopus (Brid.) Hook. f	CYRTOPODACEAE
Dactylhymenium Card. = *Didymodon*	POTTIACEAE
Dactylmenium Hag., orthogr. var. Dactylhymenium = *Didymodon*	POTTIACEAE

Genus	Family
Daltonia Hook. & Tayl., nom. cons, prop	DALTONIACEAE (Hookeriaceae)
Dasymitrium Lindb. = *Macromitrium*	ORTHOTRICHACEAE
Dawsonia R. Br	POLYTRICHACEAE (Dawsoniaceae)
Decodon C. Müll. ex Broth., nom. illeg. = *Rhachithecium*	RHACH ITHECIACEAE
Dendia R. Br. ter. = *Potlia*	POTTIACEAE
Dendralsia Hag., orthogr. var. *Dendroalsia*	CRYPHAEACEAE
Dendroalsia Britt	CRYPHAEACEAE
Dendrocryphaea Par. & Schimp. ex Broth	CRYPHAEACEAE
Dendrocyathophorum Dix	HOOKERIACEAE
Dendro-Leskea Hampe = *Porotrichum*	NECKERACEAE
Dendroligotrichum (C. Müll.) Broth	POLYTRICHACEAE
Dendropogon Schimp., hom. illeg. ≡ *Dendropogonella*	CRYPHAEACEAE
Dendropogonella Britt	CRYPHAEACEAE
Dermatodon Hüb. = p.p. *Potlia*	POTTIACEAE
Desmatodon Brid	POTTIACEAE
Desmotheca Lindb	ORTHOTRI CHACEAE
Diachelyma Payer, orthogr. var. *Dichelyma*	FONTINALACEAE
Dialytrichia (Schimp.) Limpr	POTTIACEAE
Diaphanodon Ren. & Card	TRACHYPODACEAE
Diaphanophyllum Lindb., nom. rejic. ≡ *Ditrichum*	DITRICHACEAE
Dichelodontium Hook. f. & Wils. ex Broth	PTYCHOMNIACEAE
Dichelyma Myr	FONTINALACEAE
Dichodontium Schimp	DICRANACEAE
Dicnemoloma (Ren.) Ren. = *Sclerodontium*	DICRANACEAE
Dicnemon Schwaegr	DICNEMONACEAE
Dicnemos Lindb., orthogr. var. *Dicnemon*	DICNEMONACEAE
Dicnemum Arnott, orthogr. var. *Dicnemon*	DICNEMONACEAE
Dicranella (C. Müll.) Schimp	DICRANACEAE
Dicranodon Béheré, nom. illeg. = p.p. *Dicranum* (Dicranaceae) & p.p. *Fissidens* (Fissidentaceae)	
Dicranodontium B.S.G	DICRANACEAE
Dicranoloma (Ren.) Ren	DICRANACEAE
Dicranoolontium Bruehl, orthogr. var. *Dicranodontium*	DICRANACEAE
Dicranoweisia Lindb. ex Milde	DICRANACEAE
Dicranum Hedw	DICRANACEAE
Didimodon P. Beauv., orthogr. var. *Didymodon*	POTTIACEAE
Didymodon Hedw	POTTIACEAE
Didymodon Web. & Mohr ex Braithw., hom. illeg. = *Dicranodontium*	DICRANACEAE
Didymodum P. Beauv., nom. illeg. ≡ *Didymodon*	POTTIACEAE
Dimerodontium Mitt	FABRONIACEAE
Dimorphella (C. Müll.) Ren. & Card., nom. illeg. = *Rhacopilopsis*	HYPNACEAE
Dimorphocladon Dix	HOOKERIACEAE
Diobelon Hampe, nom. illeg. = p.p. *Cynodontium*& p.p. *Dichodontium*	DICRANACEAE
Dioranum Hampe, orthogr. var. *Dicranum*	DICRANACEAE
Diphas Raf., orthogr. var. Diphascum ≡ *Diphyscium*	BUXBAUMIACEAE

Genus	Family
Diphascum A. Eat., notn. illeg. ≡ *Diphyscium*	BUXBAUMIACEAE
Diphisciam P. Beauv., orthogr. var. *Diphyscium*	BUXBAUMIACEAE
Diphyscium Mohr	BUXBAUMIACEAE (Diphysciaceae)
Diplocomium Web. & Mohr = *Meesia*	MEESIACEAE
Diploneuron Bartr	HOOKERIACEAE
Diplostichum Mont., nom. illeg. ≡ *Eustichia*	EUSTICHIACEAE
Discelium Brid	DISCELIACEAE
Discodon Hampe, orthogr. var. Dissodon = *Tayloria*	SPLACHNACEAE
Discophyllum Mitt. = *Distichophyllum*	HOOKERIACEAE
Dissodon Grev. & Arnott = *Tayloria*	SPLACHNACEAE
Distichia Brid., nom. illeg. ≡ *Neckera*	NECKERACEAE
Distichium B.S.G., nom. cons	DITRICHACEAE
Distichophyllidium Fleisch	DALTONIACEAE (Hookeriaceae)
Distichophyllum Dozy & Molk	HOOKERIACEAE
Distichum Bayrh., orthogr. var. *Distichium*	DITRICHACEAE
Ditrichium F. Müll., orthogr. var. *Ditrichum*	DITRICHACEAE
Ditrichopsis Broth	DITRICHACEAE
Ditrichum Hampe, nom. cons	DITRICHACEAE
Dixonia Horik. & Ando	LEMBOPHYLLACEAE
Dolichomitra Broth	LEMBOPHYLLACEAE
Dolichomitriopsis Okam	LEMBOPHYLLACEAE
Dolichotheca Lindb., hom. illeg. = *Herzogiella*	HYPNACEAE
Donnellia Aust. = *Sematophyllum*	SEMATOPHYLLACEAE
Dorcadion (Reichenb.) Lindb., nom. illeg. ≡ *Orthotrichum*	ORTHOTRICHACEAE
Dozya Sande Lac. in Miq	LEUCODONTACEAE
Drepanium C. Jens. = *Hypnum*	HYPNACEAE
Drepanocladus (C. Müll.) G. Roth	AMBLYSTEGIACEAE
Drepanohypnum Hampe = *Drepanocladus*	AMBLYSTEGIACEAE
Drepanophyllaria C. Müll., nom. illeg. ≡ *Cratoneuron*	AMBLYSTEGIACEAE
Drepanophyllum Schwaegr., hom. illeg. ≡ *Phyllodrepanium*	PHYLLODREPANIACEAE
Drepanopsis Loeske, nom. illeg. = *Drepanocladus*	AMBLYSTEGIACEAE
Drummondia Hook. in Drumm., nom. cons	ORTHOTRICHACEAE
Dryptodon Brid. = *Grimmia*	GRIMMIACEAE
Dubyella Schimp. = *Helicodontium*	FABRONIACEAE
Dusenia Broth., hom. illeg. ≡ *Forsstroemia*	CRYPHAEACEAE
Duseniella Broth., hom. illeg. = *Ancistrodes*	METEORIACEAE
Duthiella C. Müll. in Broth	LESKEACEAE (Trachypodaceae)
Dychelyma Croz., orthogr. var. *Dichelyma*	FONTINALACEAE
Dydimodon Arnott, orthogr. var. *Didymodon*	POTTIACEAE

Genus	Family
Eccremidium Wils	DITRICHACEAE
Echinodium Jur	ECHINODIACEAE
Ectropodon Dix. = *Anacamptodon*	FABRONIACEAE
Ectropotheciella Fleisch	HYPNACEAE
Ectropotheciopsis (Broth.) Fleisch	HYPNACEAE
Ectropothecium Mitt	HYPNACEAE

Edentella C. Müll. = *Barbula* .... POTTIACEAE
Eleutera Stuntz, nom. illeg. ≡ *Neckera* .... NECKERACEAE
Elmeriobryum Broth. (Lembophyllaceae) = *Gollania* .... RHYTIDIACEAE
Elodium (Sull.) Aust. ≡ *Helodium* .... THUIDIACEAE
Encalypta Hedw .... ENCALYPTACEAE
Encalyptra Payer, orthogr. var. *Encalypta* .... ENCALYPTACEAE
Endotrichella C. Müll. = *Garovaglia* .... PTEROBRYACEAE
Endotrichellopsis Dur .... PTEROBRYACEAE
Endotrichum Dozy & Molk., hom. illeg. =
Garovaglia .... PTEROBRYACEAE
Enthostodon auct. orthogr. var. *Entosthodon* .... FUNARIACEAE
Enlodon C. Müll .... ENTODONTACEAE
Entodontella Broth, in Fleisch .... ENTODONTACEAE
Entodontopsis Broth. = *Stereophyllum* .... PLAGIOTHECIACEAE
Entosthodon Schwaegr .... FUNARIACEAE
Entosthymenium Brid .... POTTIACEAE
Entostodon Hornsch., orthogr. var. *Entosthodon* .... FUNARIACEAE
Entothodon Brid., orthogr. var. *Entosthodon* .... FUNARIACEAE
Ephemerella C. Müll., nom. cons. = *Ephemerum* .... EPHEMERACEAE
Ephemeridium Kindb. = *Ephemerum* .... EPHEMERACEAE
Ephemeropsis Goeb .... HOOKERIACEAE (Ephemeropsaceae)
Ephemerum Hampe, nom. cons .... EPHEMERACEAE
Epipterygium Lindb .... BRYACEAE
Eremodon Brid., nom. illeg. = *Tayloria* .... SPLACHNACEAE
Eriocladium (C. Müll.) Dus., nom. illeg. =
Aerobryopsis .... METEORIACEAE
Eriodon Mont .... BRACHYTHECIACEAE
Eriopus Brid., nom. illeg. ≡ *Calyptrochaeta* .... HOOKERIACEAE
Erpodiopsis C. Müll. = *Phascum* .... POTTIACEAE
Erpodium (Brid.) Brid .... ERPODIACEAE
Erythrobarbula Steere, nom. illeg. ≡
Bryoerythrophyllum .... POTTIACEAE
Erythrodontium Hampe .... ENTODONTACEAE
Erythrophyllopsis Broth. in Herz .... POTTIACEAE
Erythrophyllum (Lindb.) Loeske, hom. illeg. ≡
Bryoerythrophyllum .... POTTIACEAE
Esenbeckia Brid., hom. illeg. ≡ *Garovaglia* .... PTEROBRYACEAE
Eucalypta Hub., orthogr. var. *Encalypta* .... ENCALYPTACEAE
Eucamptodon (Mont.) Mont .... DICNEMONACEAE
Eucamptodontopsis Broth .... DICRANACEAE
Eucatagonium Broth., nom. illeg. ≡ *Catagonium* .... PHYLLOGONIACEAE
Eucladium B.S.G .... POTTIACEAE
Euephemerum Hill., nom. illeg. ≡ *Ephemerum* .... EPHEMERACEAE
Eumyurium Nog .... MYURIACEAE
Euptichium Hampe, orthogr. var. *Euptychium* .... PTEROBRYACEAE
Euptychium Schimp .... PTEROBRYACEAE
Eurhynchiella Fleisch .... BRACHYTHECIACEAE
Eurhynchium B.S.G .... BRACHYTHECIACEAE
Eurohypnum Ando .... HYPNACEAE
Eurrhynchium B.S.G., orthogr. var. *Eurhynchium* ... BRACHYTHECIACEAE
Eurydiictyon (Card.) Hor. & Nog. =
Dendrocyathophorum .... HOOKERIACEAE
Eustichia (Brid.) Brid .... EUSTICHIACEAE

Eustichium B.S.G., nom. rejic. = *Bryoxiphium*	BRYOXIPHIACEAE
Eutodon Reim., orthogr. var. *Entodon*	ENTODONTACEAE
Euzygodon Jur., nom. illeg. ≡ *Zygodon*	ORTHOTRICHACEAE
Exodiclyon Card	CALYMPERACEAE (Leucobryaceae)
Exodokidium Card. = *Bartramia*	BARTRAMIACEAE
Fabbronia Lindb., orthogr. var. *Fabronia*	FABRONIACEAE
Fabroleskea Best = *Lindbergia*	LESKEACEAE
Fabronia Raddi	FABRONIACEAE
Fabronidium C. Müll	FABRONIACEAE
Fabroniella Lor. in C. Müll. = *Juratzkaea*	FABRONIACEAE
Farsstroemia Okam., orthogr. var. *Forsstroemia*	CRYPHAEACEAE
Fauriella Besch	THELIACEAE
Felipponea Broth. in Felipp	LEUCODONTACEAE
Fiedlera I. Hag., orthogr. var. Fiedleria = *Pterygoneurum*	POTTIACEAE
Fiedleria Rabenh., hom. illeg. = *Pterygoneurum*	POTTIACEAE
Fiorinia Schimp. in Lor., hom. illeg. = *Helicodontium*	FABRONIACEAE
Fissidens Hedw	FISSIDENTACEAE
Fissidens Lindb., hom. illeg. ≡ *Leucodon*	LEUCODONTACEAE
Fissidentella Card. = *Fissidens*	FISSIDENTACEAE
Flabellidium Herz	BRACHYTHECIACEAE
Flagellomnium Laz. = *Trachycystis*	MNIACEAE
Fleischerobryum Loeske	BARTRAMIACEAE
Florribundaria Fleisch	METEORIACEAE
Florschuetzia Crosby, nom. illeg. ≡ *Muscoftorschuetzia*	BUXBAUMIACEAE
Fontinalis Hedw	FONTINALACEAE
Foreauelia Dix. & P. de la Varde	SEMATOPHYLLACEAE
Forsstroemia Lindb	CRYPHAEACEAE
Forstroemia Par., orthogr. var. *Forsstroemia*	CRYPHAEACEAE
Fortula De Not., orthogr. var. *Tortula*	POTTIACEAE
Franciella Thér	SPIRIDENTACEAE
Fumaria Luehn., orthogr. var. *Funaria*	FUNARIACEAE
Funaria Hedw	FUNARIACEAE
X *Funariophyscomitrella* Wettst	FUNARIACEAE
Fusiconia P. Beauv.,'nom. rejic. = *Aulacomnium*	AULACOMNIACEAE
Gagea Raddi, hom. illeg. = *Zygodon*	ORTHOTRICHACEAE
Gamiella Fleisch., orthogr. var. *Gammiella*	SEMATOPHYLLACEAE
Gammiella Broth.	SEMATOPHYLLACEAE
Garckea C. Müll	DITRICHACEAE
Gorovaglia Endl	PTEROBRYACEAE
Garysmithia Steere	LESKEACEAE
Gasterogrimmia (Schimp.) Buyss. = *Grimmia*	GRIMMIACEAE
Geheebia Schimp. .	POTTIACEAE
Genthia Bayrh. ≡ *Physcomitrella*	FUNARIACEAE
Georgia C. Müll., nom. illeg. ≡ *Tetraphis*	TETRAPHIDACEAE
Gertrudia Herz., hom. illeg. ≡ *Gertrudiella*	POTTIACEAE
Gertrudiella Broth	POTTIACEAE
Gigaspermum Lindb	GIGASPERMACEAE
Giraldiella C. Müll	HYPNACEAE

Girgensobnia (Lindb.) Kindb., hom. illeg. ≡ *Pleurosiopsis* ..... PLEUROZIOPSACEAE
Glaucodipsis C. Schimp. = *Leucobryum* ..... DICRANACEAE
Globulina C. Müll., hom. illeg. ≡ *Globulinella* ..... POTTIACEAE
Globulinella Steere ..... POTTIACEAE
Glossadelphus Fleisch ..... SEMATOPHYLLACEAE
Glossophyllum (C. Müll.) Hampe, nom. illeg. = *Stereophyllum* ..... PLAGIOTHECIACEAE
Glyphocarpa R. Br. = *Bartramia* ..... BARTRAM IACEAE
Glyphocarpus Brid., nom. illeg. = *Bartramia* ..... BARTRAMIACEAE
Glyphomitrium Brid ..... PTYCHOMITRIACEAE
Glyphothecium Hampe ..... PTYCHOMNIACEAE
Glyptothecium Broth., nom. illeg. = *Glyphothecium* ... PTYCHOMNIACEAE
Gollania Broth ..... RHYTIDIACEAE
Gongronia Hag. = *Cynodontium* ..... DICRANACEAE
Goniobryum Lindb ..... RHIZOGONIACEAE
Goniomitrium Hook. & Wils ..... FUNARIACEAE
Grevillea Beck & Emmons, orthogr. var. *Grevilleanum* INSERTAESEDIS
Grevilleanum Beck & Emmons ..... INSERTAE SEDIS
Grimmia Hedw ..... GRIMMIACEAE
Groutia Broth., hom. illeg. ≡ *Dendroalsia* ..... CRYPHAEACEAE
Groutiella Steere ..... ORTHOTRICHACEAE
Guembelia Hampe in C. Müll. = *Grimmia* ..... GRIMMIACEAE
Gymbelia T. Jens., orthogr. var. Guembelia = *Grimmia* ..... GRIMMIACEAE
Gymnocephalus Schwaegr., nom. rejic. = *Aulacomnium* ..... AULACOMNIACEAE
Gymnocybe Fries, nom. rejic. = *Aulacomnium* ..... AULACOMNIACEAE
Gymnomitriella Sak. = *Gymnostomiella* ..... POTTIACEAE
Gymnostomiella Fleisch ..... POTTIACEAE (Splachnaceae)
Gymnostomum Hedw., nom. rejic.
Gymnostomum Nees & Homsch., nom. cons ..... POTTIACEAE
Gymnoweisia Mont, in Orbigny ≡ *Hymenostylium* .... POTTIACEAE
Gyrophyllum Dozy & Molk. = *Symblepharis* ..... DICRANACEAE
Gyroveisia Luis., orthogr. var. *Gyroweisia* ..... POTTIACEAE
Gyroweisia Schimp., nom. cons ..... POTTIACEAE

Habrodon Schimp ..... FABRONIACEAE
Hageniella Broth ..... SEMATOPH YL1ACEAE
Hampeella C. Müll ..... PTYCHOMNIACEAE
Handeliobryum Broth ..... NECKERACEAE
Haplocladium (C. Müll.) C. Müll ..... THUIDIACEAE
Haplodon Lindb., orthogr. var. *Aplodon* ..... SPLACHNACEAE
Haplodontium Hampe ..... BRYACEAE
Haplohymenium Dozy & Molk., nom. cons ..... THUIDIACEAE
Haplohymenium Schwaegr., nom. rejic. = *Thuidium* ..... THUIDIACEAE
Haptymenium Arnott ex Fiimr., nom. illeg. ≡ *Pterigynandrum* ..... ENTODONTACEAE
Harissona Adans, ex Leman, nom. rejic.
Harpidium (Sull.) Spruce, hom. illeg. ≡ *Drepanocladus* ..... AMBLYSTEGIACEAE

Harpophyllum Spruce ex Ren. & Card. ≡ *Hemiragis* .... HOOKERIACEAE
Harrisonia Spreng., hom. illeg. ≡ *Rhacocarpus* .... HEDWIGIACEAE
Hedvigia Kindb., orthogr. var. *Hedurigia* .... HEDWIGIACEAE
Hedwigia P. Beauv., nom. cons .... HEDWIGIACEAE
Hedwigidium B.S.G .... HEDWIGIACEAE
Helicoblepharum (Mitt.) Broth .... DALTONIACEAE (Hookeriaceae)
Helicodontiadelphus Dix .... FABRONIACEAE
Helicodontium Schwaegr .... FABRONIACEAE
Helicophyllum Brid .... HELICOPHYLLACEAE
Helmsia Bosw. = *Leptostomum* .... BRYACEAE
Helodium Wamst., nom. cons .... THUIDIACEAE
Hemiragis (Brid.) Besch .... HOOKERIACEAE
Hemisinapsium Endl., orthogr. var. Hemisynapsium = *Bryum* .... BRYACEAE
Hemisynapsium Brid. = *Bryum* .... BRYACEAE
Henediella Fleisch., orthogr. var. Hennediella = *Pottia* .... POTTIACEAE
Henicodium (C. Müll.) Kindb .... ?BRACHYTHECIACEAE
Hennedia R. Br. ter., hom. illeg. = *Pottia* .... POTTIACEAE
Hennediella Par. = *Pottia* .... POTTIACEAE
Henoniella Duby = *Ptychomitrium* .... PTYCHOMITRIACEAE
Herpetineuron (C. Müll.) Card .... THUIDIACEAE
Herpetineurum Broth., orthogr. var. *Herpetineuron* .... THUIDIACEAE
Herpodium Wittst., orthogr. var. *Erpodium* .... ERPODIACEAE
Herpodon Mont., orthogr. var. *Erpodium* .... ERPODIACEAE
Herzogiella Broth .... HYPNACEAE
Heterocladium B.S.G .... THUIDIACEAE
Heterodon Raf. = *Fissidens* .... FISSIDENTACEAE
Heterophyllium (Schimp.) Kindb. in Fleisch .... SEMATOPHYLLACEAE
Heterophyllon Kindb. ≡ *Heterophyllium* .... SEMATOPHYLLACEAE
Heterophyllum Kindb., orthogr. var. *Heterophyllium* .... SEMATOPHYLLACEAE
Hildebrandtiella C. Müll .... PTEROBRYACEAE
Himantocladium (Mitt.) Fleisch .... NECKERACEAE
Hollia Sieb. = *Dicnemon* .... DICNEMONACEAE
Holmgrenia Lindb. ≡ *Orthothecium* .... ENTODONTACEAE
Holoblepharum Dozy & Molk. = *Chaetomitrium* .... HOOKERIACEAE
Holodontium (Mitt.) Broth .... DICRANACEAE
Holomitriopsis Robins .... DICRANACEAE
Holomitrium Brid., nom. cons .... DICRANACEAE
Homalia (Brid.) B.S.G .... NECKERACEAE
Homaliadelphus Dix. & P. de la Varde .... NECKERACEAE
Homaliodendron Fleisch .... NECKERACEAE
Homaliopsis Dix. & P. de la Varde in Dix., hom. illeg. ≡ *Homaliadelphus* .... NECKERACEAE
Homalotheciella (Card.) Broth .... BRACHYTHECIACEAE
Homalothecium B.S.G .... BRACHYTHECIACEAE
Homomallium (Schimp.) Loeske .... HYPNACEAE
Hondaella Dix. & Sak .... HYPNACEAE
Hookera I. Hag., orthogr. var, *Hookeria* .... HOOKERIACEAE
Hookeria J. E. Sm., nom. cons .... HOOKERIACEAE
Hookeria Schwaegr., bom. illeg. = *Tayloria* ..., ... SPLACHNACEAE
Hookeriopsis (Besch.) Jaeg .... HOOKERIACEAE

Hookeropsis I. Hag., orthogr. var. *Hookeriopsis*	HOOKERIACEAE
Horikawaea Nog	PHYLLOGONIACEAE
Husnotiella Card	POTTIACEAE
Hyalophyllum Warnst., nom. illeg. ≡ *Stegonia*	POTTIACEAE
Hydrocryphaea Dix	NECKERACEAE (Cryphaeaceae)
Hydrogonium (C. Müll.) Jaeg	POTTIACEAE
Hydrogrimmia (I. Hag.) Loeske	GRIMMIACEAE
Hydropogon Brid	HYDROPOGONACEAE (Fontinalaceae)
Hydropogonella Card	HYDROPOGONACEAE (Fontinalaceae)
Hygroamblystegium Loeske, nom. cons	AMBLYSTEGIACEAE
Hygrddicranum Card	DICRANACEAE
Hygrohypnum Lindb	AMBLYSTEGIACEAE
Hygrypnum I. Hag., orthogr. var. *Hygrohypnum*	AMBLYSTEGIACEAE
Hylocomiastrum Fleisch. ex Broth. = *Hylocomium*	HYLOCOMIACEAE
Hylocomiopsis Card	LESKEACEAE (Thuidiaceae)
Hylocomium B.S.G	HYLOCOMIACEAE
Hymenocleiston Duby = *Tayloria*	SPLACHNACEAE
Hymenodon Hook. f. & Wils	RHIZOGONIACEAE
Hymenodontopsis Herz	RHIZOGONIACEAE
Hymenoloma Dus. (Seligeriaceae) = *Dicranoweisia*	DICRANACEAE
Hymenolomopsis Thér	SELIGERIACEAE
Hymenolopsis Bartr., orthogr. var. *Hymenolomopsis*	SELIGERIACEAE
Hymenopogon P. Beauv., orthogr. var. Hymenopogum ≡ *Diphyscium*	BUXBAUMIACEAE
Hymenopogum P. Beauv., nom. illeg. ≡ *Diphyscium*	BUXBAUMIACEAE
Hymenostoma Griff. = *Encalypta*	ENCALYPTACEAE
Hymenostomum R. Br	POTTIACEAE
Hymenostyliella Bartr	POTTIACEAE
Hymenostylium Brid	POTTIACEAE
Hymenostylium Hampe ex Garov., hom. illeg. ≡ *Amphidium*	ORTHOTRICHACEAE
Hyocomium B.S.G	HYPNACEAE
Hyophila Brid	POTTIACEAE
Hyophilopsis Card. & Dix	POTTIACEAE
Hyophilopsis Crum, hom. illeg. ≡ *N eohyophila*	POTTIACEAE
Hypnella (C. Müll.) Jaeg	SEMATOPHYLLACEAE (Hookeriaceae)
Hypnodendron (C. Müll.) Lindb. ex Mitt	HYPNODENDRACEAE
Hypnodon C. Müll. ≡ *Rhachithecium*	RHACHITHÉCIACEAE
Hypnodontopsis Iwats. & Nog	RHACHITHECIACEAE
Hypnofabronia Dix	FABRONIACEAE
Hypnopsis (Kindb.) Podp., nom. rejic. prop. ≡ *Pleurozium*	ENTODONTACEAE
Hypnum Hedw., nom. cons	HYPNACEAE
Hypodontium C. Müll	CALYMPERACEAE
Hypopterygium Brid	HOOKERIACEAE (Hypopterygiaceae)

Iisbibaea Ish, orthogr. var. Ishibaea = *Lescuraea*	LESKEACEAE
Illecebraria Hampe = *Aongstroemia*	DICBANACEAE
Indusiella Broth. & C. Müll	GRIMMIACEAE
Ischyrodon C. Müll	FABRONIACEAE
Ishibaea Broth. & Okam. (Brachytheciaceae) = *Lescuraea*	LESKEACEAE
Isocladiella Dix	PLAGIOTHECLACEAE
Isocladus Lindb. = *Sphagnum*	SPHAGNACEAE
Isodrepanium (Mitt.) Britt	NECKERACEAE (Hookeriaceae)
Isopterygiopsis Iwats	PLAGIOTHECLACEAE
Isopterygium Mitt	HYPNACEAE
Isotheciadelphus Dix. & Thér. = *Dolichomitriopsis*	LEMBOPHYLLACEAE
Isotheciopsis Broth	LEMBOPHYLLACEAE
Isothecium Brid	LEMBOPHYLLACEAE
Itatiella G. L. Sm.	POLYTRICHACEAE
Jaegerina C. Müll	PTEROBRYACEAE
Jaegerinopsis Broth. = *Jaegerina*	PTEROBRYACEAE
Jaffueliobryuni Thér	GRIMMIACEAE
Jonesia Biz., Pierr. & Pócs, hom. illeg. ≡ *Jonesiobryum*	FUNARIACEAE
Jonesiobryum Biz. & Pócs	FUNARIACEAE
Juratzkaea Lor	FABRONIACEAE
Juratzkaeella Buck	FABRONIACEAE
Kashyapia R. S. Chopra, nom. illeg. ≡ *Syrrhopodon*	CALYMPERACEAE
Kiaeria I. Hag	DICRANACEAE
Kiairia Culm., orthogr. var. *Kiaeria*	DICRANACEAE
Kibunemuscus Toyama = *Rhachithecium*	RHACHITHECIACEAE
Kingiobryum Robins	DICRANACEAE
Kisophyllum B. Pyl., orthogr. var. Skitophyllum ≡ *Fissidens*	FISSIDENTACEAE
Kleioweisiopsis Dix	POTTIACEAE
Krauseella C. Müll = *Tetraplodon*	SPLACHNACEAE
Kurohimehypnum Sak	BRACHYTHECIACEAE
	BRYACEAE
Lamprophyllum Lindb., hom. illeg. = *Pohlia*	
Lamprophyllum Schimp. ex Broth., hom. illeg. ≡	HOOKERIACEAE
Schimperobryum	CRYPHAEACEAE
Lasia P. Beauv., hom. illeg. ≡ *Forsstroemia*	ENCALYPTACEAE
Leersia Hedw. ex Batsch, hom. illeg. ≡ *Encalypta*	HYPNACEAE
Leiodontium Broth	BARTRAMIACEAE
Leiomela (Mitt.) Broth	ORTHOTRICHACEAE
Leiomitrium Mitt	ORTHOTRICHACEAE
Leiostomà (Mitt.) Par. = *Macromitrium*	ORTHOTRICHACEAE
Leiotheca Brid., nom. rejic. = *Drummondia*	
Lejotheca Hampe, orthogr. var. Leiotheca = *Drummondia*	ORTHOTRICHACEAE
Lembophyllum Lindb	LEMBOPHYLLACEAE
Lephthymenium Hag., orthogr. var. *Leptohymenium*	HYLOCOMIACEAE
Lepidopilidium (C. Müll.) Broth	HOOKERIACEAE
Lepidopilum (Brid.) Brid	DALTONIACEAE (Hookeriaceae)

Genus	Family
Leptangium Mont., nom. illeg. = *Erpodium*	ERPODIACEAE
Leptobarbula Schimp	POTTIACEAE
Leptobryum (B.S.G.) Wils	BRYACEAE
Leptochlaena Mont., hom. illeg. = *Mielichhoferia*	BRYACEAE
Leptocladiella Fleisch. = *Leptohymenium*	HYLOCOMIACEAE
Leptocladium Broth	THUIDIACEAE
Leptodictyum (Schimp.) Warnst	AMBLYSTEGIACEAE
Leptodon Mohr	NECKERACEAE
Leptodontiella Zander & E. Hege	POTTIACEAE
Leptodontiopsis Broth, in Mildbr	ORTHOTRICHACEAE
Leptodontium (C. Müll.) Hampe ex Lindb	POTTIACEAE
Leptohymenium Schwaegr	HYLOCOMIACEAE
Leptoischyrodon Dix	FABRONIACEAE
Leptopterigynandrum C. Müll	THUIDIACEAE
Leptopterygynandrum Broth., orthogr. var. *Leptopterigynandrum*	THUIDIACEAE
Leptorrhyncho-Hypnum Hampe = *Sematophyllum*	SEMATOPHYLLACEAE
Leptostomum R. Br., nom. cons	BRYACEAE (Leptostomataceae)
Leptotheca Schwaegr	AULACOMNIACEAE
Leptotrichella (C. Müll.) Lindb. = *Microdus*	DICRANACEAE
Leptotrichum Hampe ex C. Müll., hom. illeg. = *Microdus*	DITRICHACEAE
Lepyrodon Hampe	LEPYRODONTACEAE
Lepyrodontopsis Broth	BRACHYTHECIACEAE
Leratia Broth. & Par.	ORTHOTRICHACEAE
Leratiella Broth. & Syd., nom. illeg. ≡ *Leratia*	ORTHOTRICHACEAE
Lescuraea B.S.G	LESKEACEAE
Leskea Hedw	LESKEACEAE
Leskeadelphus Herz	LESKEACEAE
Leskeella (Limpr.) Loeske	LESKEACEAE
Leskeodon Broth	DALTONIACEAE (Hookeriaceae)
Leskeodontopsis Zant	DALTONIACEAE (Hookeriaceae)
Leskeopsis Broth	LESKEACEAE
Leskuraea B.S.G. ex Jaeg., orthogr. var. *Lescuraea*	LESKEACEAE
Lesquereuxia Lindb., orthogr. var. *Lescuraea*	LESKEACEAE
Lesquereuxia Lindb. ex Broth., hom. illeg. (Rhytidiaceae) = *Rigodiadelphus*	LESKEACEAE
Leucobryum Hampe	DICRANACEAE (Leucobryaceae)
Leucodon Schwaegr	LEUCODONTACEAE
Leucodoniopsis Ren. & Card., orthogr. var. *Leucodontopsis*	LEUCODONTACEAE
Leucodontella Nog. = *Leucodon*	LEUCODONTACEAE
Leucodontopsis Ren. & Card	LEUCODONTACEAE
Leucolepis Lindb	MNIACEAE
Leucoloma Brid., nom. cons	DICRANACEAE
Leucomium Mitt	HYPNACEAE (Leucomiaceae)
Leucomnium Mitt., orthogr. var. *Leucomium*	HYPNACEAE
Leucophanella (Besch.) Fleisch. ≡ *Syrrhopodon*	CALYMPERACEAE

Leucophanes Brid .... CALVMPERACEAE (Leucobryaceae)
Levierella C. Müll .... ENTODONTACEAE
Ligulina C. Müll. = p.p. *Plagiothecium* (Plagiotheciaceae) & p.p. *Potamium* (Sematophyllaceae)
Limbella (Broth.) Bartr .... PTEROBRYACEAE
Limneria Stirt. = *Didymodon* .... POTTIACEAE
Limnobiella (C. Müll.) C. Müll. = *Glossadelphus* .... SEMATOPHYLLACEAE
Limnobium B.S.G., hom. illeg. ≡ *Hygrohypnum* .... AMBLYSTEGIACEAE
Limnobryum Rabenh. = *Aulacomnium* .... AULACOMNIACEAE
Limprichtia Loeske = *Drepanocladus* .... AMBLYSTEGIACEAE
Lindbergia Kindb .... LESKEACEAE
Lindigia Hampe .... METEORIACEAE B
Liomela Hag., orthogr. var. *Leiomela* .... ARTRAMIACEAE
Liomitrium Hag., orthogr. var. *Leiomitrium* .... ORTHOTRICHACEAE
Liopilus Mohr ex Web .... INCERTAE SEDIS
Liotheca Hag., orthogr. var. Leiotheca = *Drummondia* .... ORTHOTRICHACEAE
Loeskea Fleisch., orthogr. var. *Leskea* .... LESKEACEAE
Loeskeobryum Fleisch. in Broth. (Rhytidiaceae) = *Hylocomium* .... HYLOCOMIACEAE
Loeskypnum Paul = *Drepanocladus* .... AMBLYSTEGIACEAE
Loiseaubryum Biz .... FUNARIACEAE
Lomoporotrichum C. Müll. = *Sciaromium* .... AMBLYSTEGIACEAE
Lophidium Rodw., orthogr. var. *Lopidium* .... HOOKERIACEAE
Lophiodon Hook. f. & Wils., nom. rejic. = *Ditrichum* .... DITRICHACEAE
Lopidium Hook. f. & Wils .... HOOKERIACEAE (Hypopterygiaceae)

Lorentzia Hampe = *Pelekium* .... THUIDIACEAE
Loreniziella C. Müll .... GIGASPERMACEAE
Luisicrella Thér. & P. de la Varde .... POTTIACEAE
Lydiaea Laz. = *Phascum* .... POTTIACEAE
Lyellia R. Br .... POLYTRICHACEAE

Macgregorella Bartr. = *Schwetschkea* .... FABRONIACEAE
Macouniella Kindb. = *Antitrichia* .... LEUCODONTACEAE
Macrauchenangium Wijk, Marg. & Florsch., orthogr. var. Macrauchenium Brid. = p.p. *Plagiobryum* & p.p. *Pohlia* .... BRYACEAE
Macrauchenium Brid. = p.p. *Plagiobryum* & p.p. *Pohlia* .... BRYACEAE
Macrocoma (C. Müll.) Grout .... ORTHOTRICHACEAE
Macrodon Arnott, nom. rejic. = *Leucoloma* .... DICRANACEAE
Macroglossum Hilp., nom. illeg. = *Hyophila* .... POTTIACEAE
Macrohymenium C. Müll .... SEMATOPHYLLACEAE
Macrohymenium C. Müll, in Doty & Molk., hom. illeg .... REGMATODONTACEAE
Macromiirium Brid .... ORTHOTRICHACEAE
Macrosporiella Dix. & Thér. = *Leucodon* .... LEUCODONTACEAE
Macrothamniella Fleisch .... HYLOCOMIACEAE
Macrothamnium Fleisch .... HYLOCOMIACEAE

Macrothecium Brid., orthogr. var. Megalangium ≡ *Acidodontium*	BRYACEAE
Macrymenium Hag., Orthogr. var. Macrohymeniutn C. Müll, in Dozy & Molk., hom. illeg	REGMATODONTACEAE
Maireola Thér. & Trab	DICRANACEAE
Mamillariella Laz	FABRONIACEAE
Mandoniella Herz	BRACHYTHECIACEAE
Mascalomthus Lem., orthogr. var. Maschalanthus = *Pterigynandrum*	ENTODONTACEAE
Maschalanthus Spreng, ex K. F. Schultz, nom. illeg. = *Pterigynandrum*	ENTODONTACEAE
Maschalocarpus Spreng., nom. illeg. ≡ *Pterigynandrum*	ENTODONTACEAE
Mastopoma Card	SEMATOPHYLLACEAE
Matsumuraea Okam. = *Duthiella*	LESKEACEAE
Meesea C. Müll., orthogr. var. *Meesia*	MEESIACEAE
Meesia Hedw., nom. cons	MEESIACEAE
Megalangium Brid., nom. rejic. ≡ *Acidodontium*	BRYACEAE
Meiotheciopsis Broth	SEMATOPHYLLACEAE
Meiothccium Mitt	SEMATOPHYLLACEAE
Melophyllum Herz. = *Bryomanginia*	DITRICHACEAE
Merceya Schimp. = *Scopelophila*	POTTIACEAE
Merceyopsis Broth. & Dix	POTTIACEAE
Merrilltobryum Broth	FABRONIACEAE
Mesochaete Lindb	RHIZOGONIACEAE
Mesonodon Hampe	ENTODONTACEAE
Mesotus Mitt, in Hook, f	DICNEMONACEAE
Metadistichophyllum Nog. & Iwats. = *Distichophyllum*	HOOKERIACEAE
Metaneckera Steere	NECKERACEAE
Meteoridium (C. Müll.) Manuel	METEORIACEAE
Meteoriella Okam	METEORIACEAE
Meteoriopsis Fleisch. ex Broth	METEORIACEAE
Meteorium (Brid.) Dozy & Molk	METEORIACEAE
Metorium Reichenb., orthogr. var. *Meteorium*	METEORIACEAE
Metzlera Hag., orthogr. var. Metzleria = *Atractylocarpus*	DICRANACEAE
Metzlerella Hag. = *Atractylocarpus*	DICRANACEAE
Metzleria Schimp. ex Milde, nom. rejic. = *Atractylocarpus*	DICRANACEAE
Microbryum Schimp. = *Phascum*	POTTIACEAE
Microcampylopus (C. Müll.) Fleisch	DICRANACEAE
Microctenidium Fleisch	HYPNACEAE
Microdendron Broth	POLYTRICHACEAE
Microdus Schimp. in Besch	DICRANACEAE
Micromitrium Aust	EPHEMERACEAE
Micromitrium Schimp. in Besch., hom. illeg. ≡ *Groutiella*	ORTHOTRICHACEAE
Microphilonotis Sak. = *Pohlia*	BRYACEAE
Micropoma Lindb	FUNARIACEAE
Microstegium Lindb., hom. illeg. ≡ *Micropoma*	FUNARIACEAE
Microthamnium Mitt., hom. illeg. ≡ *Mittenothamnium*	HYPNACEAE
Microtheciella Dix. (Erpodiaceae)	MICROTHECIELLACEAE

Microthuidium (Limpr.) Warnst. = *Thuidium* .... THUIDIACEAE
Mielichhofera Hag., orthogr. var. *Mielichhoferia* .... BRYACEAE
Mielichhoferia Nees & Homsch .... BRYACEAE
Mielichoferia Spruce, orthogr. var. *Mielichhoferio* .... BRYACEAE
Mildea Warnst., hom. illeg. = *Pottia* .... POTTIACEAE
Mildeella Limpr., hom. illeg. = *Pottia* .... POTTIACEAE
Miothecium Mitt., orthogr. var. *Meiothecium* .... SEMATOPHYLLACEAE
Mitrapoma Duby = *Calyptrochaeta* .... HOOKERIACEAE
Mitrobryum Robins .... DICRANACEAE
Mitropoma Duby, orthogr. var. Mitrapoma = *Calyptrochaeta* .... HOOKERIACEAE
Mittenia Lindb .... MITTENIACEAE
Mittenothamnium Henn., nom. cons .... HYPNACEAE
Mitthyridium Robins .... CALYMPERACEAE
Miyabea Broth .... THUIDIACEAE
Mniadelphus C. Müll., nom. illeg. ≡ *Distichophyllum* .... HOOKERIACEAE
Mniobryoïdes Hörm .... BRYACEAE
Mniobryum Limpr., nom. cons .... BRYACEAE
Mniodendron Lindb. in Dozy & Molk. = *Hypnodendron* .... HYPNODENDRACEAE
Mniodendrum Hag., orthogr. var. Mniodendron = *Hypnodendron* .... HYPNODENDRACEAE
Mniomalia C. Müll .... PHYLLODREPANIACEAE
Mniomallia Bartr., orthogr. var. *Mniomalia* .... PHYLLODREPANIACEAE
Mniopsis Mitt, in Hook, f., hom. illeg. ≡ *Mittenia* .... MITTENIACEAE
Mnium Hedw., nom. cons .... MNIACEAE
Moenckemeyera Broth., orthogr. var. Moenkemeyera = *Fissidens* .... FISSIDENTACEAE
Moenkemeyera C. Müll. = *Fissidens* .... FISSIDENTACEAE
Molendoa Lindb .... POTTIACEAE
Mollia Schrank, ex Lindb., hom. illeg. = p.p. *Barbula* & p.p. *Tortula* .... POTTIACEAE
*Monocranum*C. Müll .... DICRANACEAE
Monoschisma Duby = *Meteorium* .... METEORIACEAE
Morinia Card .... POTTIACEAE
Moritzia Hampe, hom. illeg. = *Schoenobryum* .... CRYPHAEACEAE
Moseniella Broth .... SPLACHNACEAE
Muelleriella Dus , nom. cons .... ORTHOTRICHACEAE
Muellerobryum Fleisch .... PTEROBRYACEAE
Muscoflorschuetzia Crosby .... BUXBAUMIACEAE
Myrinia Schimp., nom. cons .... FABRONLACEAE
Myurella B.S.G .... THFLIACEAE
Myuriopsis Nog. ≡ *Eumyurium* .... MYURIACEAE
Myurium Schimp .... MYURIACEAE
Myuroclada Besch .... BRACHYTHECIACEAE

Nadeaudia Besch. ≡ *Calomnion* .... CALOMNIACEAE
Nanobryum Dix .... NANOBRYACEAE (Dicranaceae)
Nonomitriella Bartr .... FUNARIACEAE
Nanomitriopsis Card. (Ephemeraceae) = *Pseudephemerum* .... DICRANACEAE
Nanomitrium Lindb., nom. illeg. = *Micromitrium* .... EPHEMERACEAE

Nanothecium Dix. & P. de la Varde	ENTODONTACEAE
Neacroporium Iwats. & Nog	SEMATOPHYLLACEAE
Neckera Hedw., nom. cons	NECKERACEAE
Neckeradelphus Laz. = *Neckera*	NECKERACEAE
Neckeradelphus Steere, hom. illeg. ≡ *Metaneckera*	NECKERACEAE
Neckeropsis Reichardt	NECKERACEAE
Neobarbella Nog. = *Isotheciopsis*	LEMBOPHYLLACEAE
Neobarbula Dus. = *Tortula*	POTTIACEAE
Neobryum Williams = *Osculatia*	BRYACEAE
Neocardotia Thér, & Bartr. = *Rhexophyllum*	POTTIACEAE
Neodolichomitra Nog	RHYTIDIACEAE
Neohyophila Crum	POTTIACEAE
Neohypnella Bartr	SEMATOPHYLLACEAE
Neolescuraea Nog. = *Rigodiadelphus*	LESKEACEAE
Neolindbergia Fleisch	PTEROBRYACEAE (Prionodontaceae)
Neoloma Card., orthogr. var. Neuroloma = *Andreaea*	ANDREAEACEAE
Neomacounia Irel	NECKERACEAE
Neorutenbergia Biz. & Pócs	RUTENBERCIACEAE
Neosharpiella Robins. & Delg	FUNARIACEAE
Neuroloma Card., hom. illeg. = *Andreaea*	ANDREAEACEAE
Notarisia Hampe, hom. illeg. = *Ptychomitrium*	PTYCHOMITRIACEAE
Notoligotrichum G. L. Sm	POLYTRICHACEAE
Nyholmiella Holmen & Wamcke = *Orthotrichum*	ORTHOTRICHACEAE
Ochrobryum Mitt	DICRANACEAE (Leucobryaceae)
Octoblepharis Lindb., orthogr. var. *Octoblepharum*	CALYMPERACEAE
Octoblepharum Hedw	CALYMPERACEAE (Leucobryaceae)
Octodiceras Brid. = *Fissidens*	FISSIDENTACEAE
Octogonella Dix	ORHOTRICHACEAE
Oedicladium Mitt. = *Myurium*	MYURIACEAE
Oedipodiella Dix	GIGASPERMACEAE
Oedipodium Schwaegr	OEDIPODIACEAE
Okamuraea Broth	RHYTIDIACEAE
Oligotrichum Lam. & Cand	POLYTRICHACEAE
Olomitrium Brid., nom. rejic. ≡ *Holomitrium*	DICRANACEAE
Omalia (Brid) B.S.G., orthogr. var. *Homalia*	NECKERACEAE
Oncophora Wallr., nom. illeg. ≡ *Oncophorus*	DICRANACEAE
Oncophorus (Brid.) Brid	DICRANACEAE
Oreas Brid	DICRANACEAE
Oreas Brid, in Lindb., hom. illeg. ≡ *Mieiichhoferia*	BRYACEAE
Oreoweisia (B.S.G.) De Not	DICRANACEAE
Orontobryum Mitt, ex Fleisch	ENTODONTACEAE (Hookeriaceae)
Orthoamblystegium Dix. & Sak	LESKEACEAE
Orthodicranum (B.S.G.) Loeske	DICRANACEAE
Orthodon R. Br. = *Tayloria*	SPLACHNACEAE
Orthodontium Schwaegr	BRYACEAE
Ortholimnobium Dix	AMBLYSTEGIACEAE
Orthomnion Wils	MNIACEAE

Orthomniopsis Broth ... MNIACEAE
Orthomnium Broth., orthogr. var. Orthomnion ... MNIACEAE
Orthopixis P. Beauv., orthogr. var. Orthopyxis ≡
Aulacomnium ... AULACOMNIACEAE
Orthopus Wulfsb. = Campylopus ... DICRANACEAE
Orthopyxis P. Beauv., nom. rejic. ≡
Aulacomnium ... AULACOMNIACEAE
Orthorhynchium Reichardt, orthogr. var.
Orthorrhynchium ... PHYLLOGONIACEAE
Orthorrhynchidium Ren. & Card ... PTEROBRYACEAE
Orthorrhynchium Reichardt ... PHYLLOGONIACEAE
Orthostichella C. Müll. = *Pilotrichella* ... METEORIACEAE
Orthostichidium C. Müll. ex Dus ... PTEROBRYACEAE
Orthostichopsis Broth ... PTEROBRYACEAE
Orthotheca Brid. = Syrrhopodon ... CALYMPERACEAE
Orthotheciadelphus Dix ... ENTODONTACEAE
Orthothecium B.S.G., nom. cons ... ENTODONTACEAE
Orthotrichum Hedw ... ORTHOTRICHACEAE
Osculatia De Not ... BRYACEAE
Osmundula Rabenh. = Fissidens ... FISSIDENTACEAE
Osterwaldiella Fleisch. ex Broth ... PTEROBRYACEAE
Oticodium (C. Müll.) Kindb. = Palamocladium ... BRACHYTHECIACEAE
Oxyrrhynchium (B.S.G.) Wamst ... BRACHYTHECIACEAE
Oxystegus (Limpr.) Hilp ... POTTIACEAE

Pachyfissidens (C. Müll.) Limpr. = Fissidens ... FISSIDENTACEAE
Pachyneuropsis Mill ... POTTIACEAE
Pachyneurum Amann = Tortula ... POTTIACEAE
Pachyneurum Bartr., hom. illeg. ≡ Pachyneuropsis ... POTTIACEAE
Palamocladium C. Müll. ... BRACHYTHECIACEAE
Palisadula Toyama ... SEMATOPHYLLACEAE
Paludella Ehrh. ex Brid ... MEESIACEAE
Panckowia (Lindb.) Monk., hom. illeg. =
Eurhynchium ... BRACHYTHECIACEAE
Pancovia Neck, ex Kickx, = Eurhynchium ... BRACHYTHECIACEAE
Papillaria (C. Müll.) C. Müll., nom. cons ... METEORIACEAE
Paraleucobryum (Limpr.) Loeske ... DICRANACEAE
Paramyurium (Limpr.) Warnst. = Cirriphyllum ... BRACHYTHECIACEAE
Parisia Broth ... DICRANACEAE
Pelecium Hag., orthogr. var. Pelekium ... THUIDIACEAE
Pelekium Mitt ... THUIDIACEAE
Pentastichella C. Müll. = Stenomitrium ... ORTHOTRICHACEAE
Penzigiella Fleisch ... PTEROBRYACEAE
Perisiphorus P. Beauv. = Hedwigia ... HEDWIGIACEAE
Peromnion Schwaegr. = Brachymenium ... BRYACEAE
Perssonia Biz ... BRYACEAE
Pharomitrium Schimp., nom. rejic. = Pterygoneurum POTTIACEAE
Phasconica C. Müll ... POTTIACEAE
Phascum Hedw ... POTTIACEAE
Philibertiella Card ... DITRICHACEAE
Philocrya Hag. & C. Jens. = Lyeilia ... POLYTRICHACEAE
Philonotis Brid ... BARTRAMIACEAE
Philonotula (B.S.G.) B.S.G. = Philonotis ... BARTKAMIACEAE

Genus	Family
Philonotula Hampe, hom. illeg. = *Bartramidula*	BARTRAMIACEAE
Philophyllum C. Müll	HOOKERIACEAE
Philoscia Berk. = *Plagiothecium*	PLAGIOTHECIACEAE
Pblogostomum Hampe, nom. illeg. = *Clasmatodon*....................	FABRONIACEAE
Photinophyllum Mitt., nom. illeg. = *Goniobryum*......................	RHIZOGONIACEAE
Phyllodon B.S.G	HOOKERIACEAE
Phyllodrepanium Crosby	PHYLLODREPANIACEAE
Phyllogonium Brid.	PHYLLOGONIACEAE
Phylonotis Papp, orthogr. var. *Philonotis*	BARTRAMIACEAE
Physcomitrella B.S.G	FUNARIACEAE
Physcomitrellopsis Broth. & Wager ex Dix	FUNARIACEAE
Physcomitridium (C Müll.) G. Roth = *Physcomitrella*	FUNARIACEAE
Physcomitrium (Brid.) Brid	FUNARIACEAE
Physedium Brid. = *Pottia*..	POTTIACEAE
Physodictyon Kuetz	INSERTAE SEDIS
Pilaisaea Desv. in B. Pyl. = *Plagiothecium*................................	PLAGIOTHECIACEAE
Pilaisia Hérib., orthogr. var. *Pylaisia*	HYPNACEAE
Piloecium (C. Müll.) Broth	MYURIACEAE
Pilopogon Brid	DICRANACEAE
Pilopogonella Bartr	DICRANACEAE
Pilosium (C. Müll.) Fleisch.......................................	PLAGIOTHECIACEAE
Pilotrichella (C. Müll.) Besch..	METEORIACEAE
Pilotrichidium Besch	HOOKERIACEAE (Pilotrichaceae)
Pilotrichopsis Besch	CRYPHAEACEAE
Pilotrichum P. Beauv. = *Leptodon*	NECKERACEAE
Pilotrichum Brid., hom. illeg. (Pilotrichaceae) = *Callicosta*	DALTONIACEAE
Pinnatella Fleisch	NECKERACEAE
Pinnatera Okam., orthogr. var. *Pinnatella*	NECKERACEAE
Pirea Card., hom. illeg. ≡ *Pireella*	PTEROBRYACEAE
Pireella Card................................ , ...	PTEROBRYACEAE
Plagiobryum Lindb	BRYACEAE
Plagiomnium T. Kop...	MNIACEAE
Plagiopus Brid	BARTRAMIACEAE
Plagiotheciella Fleisch. ex Broth., nom. illeg. (Hypnaceae) = *Plagiothecium*...	PLAGIOTHECIACEAE
Plagiotheciopsis Broth..	ENTODONTACEAE
Plagiothecium B.S.G., nom. cons. prop	PLAGIOTHECIACEAE
Plasteurhynchium Fleisch. ex Broth	LEMBOPHYLLACEAE
Platihypnidium Sak., orthogr. var. *Platyhypnidium*	AMBLYSTEGIACEAE
Platydictya Berk	AMBLYSTEGIACEAE
Platygyriella Card...	HYPNACEAE
Platygyrium B.S.G., nom. cons..	HYPNACEAE
Platyhypnidium Fleisch...	AMBLYSTEGIACEAE
Platyhypnum Loeske = *Hygrohypnum*	AMBLYSTEGIACEAE
Platyloma Kindb., hom. illeg. = *Platylomella*...........................	AMBLYSTEGIACEAE
Platylomella Andrews..	AMBLYSTEGIACEAE
Platyneurum (Card.) Broth	DICRANACEAE
Plaubelia Brid., nom. rejic. = *Desmatodon*	POTTIACEAE
X *Pleuriditrichum* Andrews & Hermann	DITRICHACEAE
Pleuridiella Robins	DITRICHACEAE
Pleuridium Rabenh., nom. cons	DITRICHACEAE

Pleuridium Brid., nom. rejic. = *Archidium* .... ARCHIDIACEAE
Pleurochaete Lindb .... POTTIACEAE
Pleurophascum Lindb .... PLEUROPHASCACEAE
Pleuropus Griff., hom. illeg. ≡ *Palamocladium* .... BRACHYTHECIACEAE
Pleurorthotrichum Broth .... - .... ORTHOTRICHACEAE
Pleuroweisia Limpr .... POTTIACEAE
Pleuroziopsis Kindb. ex Britt .... PLEUROZIOPSACEAE (Climaciaceae)
Pleurozium Mitt., nom. cons, prop .... ENTODONTACEAE
Pleurozium (Sull.) Mitt., nom. illeg. ≡ *Hylocomium* .... HYLOCOMIACEAE
Pleurozygodon Lindb. = *Anoectangium* .... POTTIACEAE
Pleurozygodontopsis Dix .... ORTHOTRICHACEAE
Poecilophyllum Mitt. = *Leucoloma* .... DICRANACEAE
Pogonatum P. Beauv .... POLYTRICHACEAE
Pohlia Hedw .... BRYACEAE
Polla (Brid.) Loeske ≡*Mnium* .... MNIACEAE
Polymerodon Herz .... DICRANACEAE
Polypodiopsis (C. Müll.) Jaeg. = *Fissidens* .... FISSIDENTACEAE
Polytrichadelphus (C. Müll.) Mitt .... POLYTRICHACEAE
Polytrichastrum G. L. Sm. .... POLYTRICHACEAE
Polytrichodendron Fleisch. = *Dendroligotrichum* .... POLYTRICHACEAE
Polytrichum Hedw .... POLYTRICHACEAE
Parothamnium Fleisch .... NECKERACEAE
Porotrichella Card, in Grand. = *Qrthostichopsis* .... PTEROBRYACEAE
Porotrichodendron Fleisch .... LEMBOPHYLLACEAE
Porotrichodendrum Fleisch., orthogr. var. *Porotrichodendron* .... LEMBOPHYLLACEAE
Porotrichopsis Broth. & Herz .... LEMBOPHYLLACEAE
Porotrichum (Brid.) Hampe .... NECKERACEAE
Potamium Mitt .... SEMATOPHYLLACEAE
Pottia (Reichenb.) Fürnr .... POTTIACEAE
Pottiella (Limpr.) Gams = *Phascum* .... POTTIACEAE
Powellia Mitt .... RACOPILACEAE
Pringleella Card .... DITRICHACEAE
Pringleella Broth., hom. illeg. = *Pringleella* Card. .... DITRICHACEAE
Prionidium Hilp .... POTTIACEAE
Prionodon C. Müll .... PRIONODONTACEAE
Pseudatrichum Reim .... POLYTRICHACEAE
Pseudephemerum (Lindb.) I. Hag. .... DICRANACEAE
Pseudisothecium Grout = *Isothecium* .... LEMBOPHYLLACEAE
Pseudobarbella Nog .... METEORIACEAE
Pseudobraunia (Lesq. & James) Broth .... HEDWIGIACEAE
Pseudobryum (Kindb.) T. Kop .... MNIACEAE
Pseudocalliergon (Limpr.) Loeske = p.p. *Calliergon & p.p. Drepanocladus* .... AMBLYSTEGIACEAE
Pseudocrossidium Williams .... POTTIACEAE
Pseudocryphaea Brid. ex Broth .... LEUCODONTACEAE
Pseudodimerodontium (Broth.) Broth .... FABRONIACEAE
Pseudodistichium Card .... DITRICHACEAE
Pseudoditrichum Steere & Iwats .... PSEUDODITRICHACEAE
Pseudoephemerum I. Hag., orthogr. var. *Pseudephermerum* .... DICRANACEAE
Pseudohygrohypnum Kanda .... AMBLYSTEGIACEAE
Pseudohyophila Hilp .... GRIMMIACEAE

Pseudohypnella (Broth.) Fleisch .... SEMATOPHYLLACEAE (Hookeriaceae)
Pseudoisothecium Grout *ex* Podp , orthogr. var. Pseudisothecium = *Isothecium* .... LEMBOPHYLLACEAE
Pseudoleskea B.S.G. .... LESKEACEAE
Pseudoleskeella Kindb .... LESKEACEAE
Pseudoleskeopsis Broth .... LESKEACEAE
Pseudopiloecium Bartr .... SEMATOPHYLLACEAE
Pseudopleuropus Tak .... BRACHYTHECIACEAE
Pseudopohlia Williams .... BRYACEAE
Pseudopterobryum Broth .... PTEROBRYACEAE
Pseudoracelopus Broth .... POLYTRICHACEAE
Pseudorhacelopus Broth., orthogr. var. *Pseudoracelopus* .... POLYTRICHACEAE
Pseudoscleropodium (Limpr.) Fleisch. ex Broth .... ENTODONTACEAE
Pseudospiridentopsis (Broth.) Fleisch .... TRACHYPODACEAE
Pseudostereodon (Broth.) Fleisch .... HYPNACEAE
Pseudosymblepharis Broth .... POTTIACEAE
Pseudoterobryum Wijk, Marg. & Florsch., orthogr. var. *Pseudopterobryum* .... PTEROBRYACEAE
Pseudothuidium Herz. = *Diaphanodon* .... TRACHYPODACEAE
Pseudotrachypus P. de la Varde & Thér .... TRACHYPODACEAE
Pseudoweisia Krause, nom. illeg. ≡ *Dicranoweisia* .... DICRANACEAE
Psilopilum Brid .... POLYTRICHACEAE
Pterigoneurum Jur., orthogr. var. ≡ *Pterygoneurum* .... POTTIACEAE
Pterigonium Gray, orthogr. var. *Pterogonium* .... LEUCODONTACEAE
Pterigpphyllum Brid., orthogr. var. Pterygophyllum Brid. ≡ *Hookeria* .... HOOKERIACEAE
Plerigynandrum Hedw .... ENTODONTACEAE
Pterobryella (C. Müll.) Jaeg .... PTEROBRYACEAE
Pterobryidium Broth. & Watts .... PTEROBRYACEAE
Pterobryon Hornsch .... PTEROBRYACEAE
Pterobryopsis Fleisch .... PTEROBRYACEAE
Pterobryum Hornsch., orthogr. var. *Pterobryon* .... PTEROBRYACEAE
Pterogoneurum Jur., orthogr. var. *Pterygoneurum* .... POTTIACEAE
Pterogoniadelphus Fleisch. = *Filipponea* .... LEUCODONTACKAE
Pterogonidium C. Müll. ex Broth .... SEMATOPHYLLACEAE
Pterogoniella Schimp. ex Broth. = *Meiothecium* .... SEMATOPHYLLACEAE
Pterogoniopsis C. Müll .... SEMATOPHYLLACEAE
Pterogonium Sw .... LEUCODONTACEAE
Pterogynandrum Herib., orthogr. var. *Pterigyandrum* .... ENTODONTACEAE
Pterygoneuron Kindb., orthogr. var. *Pterygoneurum* .... POTTIACEAE
Pterygoneurum Jur., nom. cons .... POTTIACEAE
Pterygophyllum Brid., nom. illeg. ≡ *Hookeria* .... HOOKERIACEAE
Pterygophyllum Broth., hom. illeg. ≡ *Achrophyllum* .... HOOKERIACEAE
Pterygynandrum Hedw., orthogr. var. *Pterigynandrum* .... ENTODONTACEAE
Pthychomitrium Fürnr., nom. rejic. = *Ptychomitrium* .... PTYCHOMITRIACEAE
Pticodes Lem., orthogr. var. Ptychodes ≡ *Schlotheimia* .... ORTHOTRICHACEAE
Ptichodium Pavl., orthogr. var. Ptychodium = *Lescuraea* .... LESKEACEAE
Pticomitrium Fürnr., orthogr. var. *Ptychomitrium* .... PTYCHOMITRIACEAE
Ptilium De Not .... HYPNACEAE

Ptilocladus Lindb. = *Campiochaete*.. LEMBOPHYLLACEAE
Ptychodes Web. ≡ *Schlotheimia* .... ORTHOTRICHACEAE
Ptychodium Schimp. (Rhytidiaceae) = *Lescuraea* .... LESKEACEAE
Ptychomitriopsis Dix.. PTYCHOMITRIACEAE
Plychomitrium Fürnr., nom. cons .... PTYCHOMITRIACEAE
Ptychomnion (Hook. f. & Wils.) Mitt.. PTYCHOMNIACEAE
Ptychomnium Broth., orthogr. var. *Ptychomnion* .... PTYCHOMNIACEAE
Ptychophyllum Broth., hom. illeg.. HYPNACEAE
Ptychophyllum Dix. &Herz., hom. illeg. ≡ Ptychophyllum
Broth., hom. illeg .... HYPNACEAE
Ptychostomum Hornsch. = *Bryum*... BRYACEAE
Ptycomitrium Angstr., orthogr. var. *Ptychomitrium* PTYCHOMITRIACEAE
Puiggaria Duby = *Lepidopilum* .... DALTONIACEAE
Puiggariella Broth.. HYPNACEAE
Puiggarria Wijk, Marg. & Florsch., orthogr. var.
Puiggaria = *Lepidopilum* .... DALTONIACEAE
Pulvinella Broth. & Herz.. HOOKERIACEAE
Pungentella C. Müll .... SEMATOPHYLLACEAE
Pycnapophyscium (Reichenb.) C. Müll., nom. illeg. ≡
Tetraplodon................................. SPEACHNACEAE
Pylaiea Lindb., orthogr. var. *Pylaisia* .... HYPNACEAE
Pylaisaea B. Pyl., orthogr. var. Pilaisaea =
Plagiothecium .... PLAGIOTHECIACEAE
Pylaisaea B.S.G., hom. illeg. = *Pylaisia*.................................... HYPNACEAE
Pylaisea Fürnr., orthogr. var. Pilaisaea =
Plagiothecium .... PLAGIOTHECIACEAE
Pylaisia Schimp.. HYPNACEAE
Pylaisiadelpha Card .... SEMATOPHYLLACEAE
Pylaisiella Kindb. ex Grout = *Pylaisia* .... HYPNACEAE
Pylaisiobryum Broth.. ENTODONTACEAE
Pylaisiopsis Broth .... SEMATOPHYLLACEAE
Pyloecium Par. ≡ *Piloecium* .... MYURIACEAE
Pyramidium Brid., nom. illeg. ≡ *Pyramidula* .... FUNARIACEAE
Pyramidula Brid.. FUNARIACEAE
Pyramitrium Mitt. = *Encalypta*... ENCALYPTACEAE
Pyromitrium (Hampe) Wallr. ex Hampe =
Encalypta .... ENCALYPTACEAE
Pyrrhobryum Mitt. = *Rhizogonium*.. RHIZOGONIACEAE

Racelopodopsis Thér .... POLYTRICHACEAE
Racelopus Dozy & Molk.. POLYTRICHACEAE
Racocarpus Kindb., orthogr. var. *Rhacocarpus*.......................... HEDWIGIACEAE
Racomitrium Brid.. GRIMMIACEAE
Racopilum P. Beauv .... RACOPILACEAE
Raine ria De Not = *Tayloria*.. SPLACHNACEAE
Raphidostegium (B.S.G.) Besch., orthogr. var.
Rhaphidostegium ≡ *Sematophyllum*.. SEMATOPHYLLACEAE
Ratthynidium Wijk, Marg. & Florsch., orthogr. var.
Platyhypnidium.. AMBLYSTEGIACEAE
Ratyhypnidium Bruehl, orthogr. var. *Platyhypnidium* AMBLYSTEGIACEAE
Raua I. Hag., orthogr. var. Rauia ≡ *Rauiella* .... THUIDIACEAE
Rauia Aust., hom. illeg. ≡ *Rauiella* .... THUIDIACEAE
Rauiella Reim.. THUIDIACEAE
Rechingerella J. Fröhl. = *Weissia* .... POTTIACEAE

Genus	Family
Regmatodon Brid	REGMATODONTACEAE (Leskeaceae)
Rehmanniella C. Müll. = *Goniomitrium*	FUNARIACEAE
Reimersia Chen	POTTIACEAE
Remyella C. Müll. = *Rhynchostegiella*	BRACHYTHECLACEAE
Renauldia C. Müll. in Ren	PTEROBRYACEAE
Reddens Dix	ENTODONTACEAE
Rhabdodontium Broth	PTEROBRYACEAE
Rhabdoweisia B.S.G	RHABDOWEISIACEAE (Dicranaceae)
Rhabdoweisiella Williams	RHABDOWEISIACEAE (Dicranaceae)
Rhabdoweissia Mont., orthogr. var. *Rhabdoweisia*	RHABDOWEISIACEAE
Rhacelopodopsis Broth., orthogr. var. *Racelopodopsis*	POLYTRICHACEAE
Rhacelopus Sande Lac., orthogr. var. *Racelopus*	POLYTRICHACEAE
Rhachitheciopsis P. de la Varde = *Rhachithecium*	RHACHITHECIACEAE
Rhachithecium Broth, ex Le Jol.	RHACHITHECIACEAE (Orthotrichaceae)
Rhacocarpus Lindb	HEDWIGIACEAE
Rhacomitrium Brid., orthogr. var. *Racomitrium*	GRIMMIACEAE
Rhacopilopsis Ren. & Card	HYPNACEAE
Rhacopilum Mitt., orthogr. var. *Racopilum*	RACOPILACEAE
Rhamphidium Mitt	POTTIACEAE
Rhaphidorrhynchium Fleisch. = *Pungentella*	SEMATOPHYLLACEAE
Rhapidostegium (B.S.G.) De Not., nom. illeg. ≡ *Semaiophyllum*	SEMATOPHYLLACEAE
Rhapidosdchum Fleisch	SEMATOPHYLLACEAE
Rbegmatodon Reichenb., orthogr. var. *Regmatodon*	REGMATODONTACEAB
Rhexophyllum Herz	POTTIACEAE
Rhincostegium Anzi, orthogr. var. *Rhynchostegium*	BRACHYTHECLACEAE
Rhizofabronia (Broth.) Fleisch	FABRONIACEAE
Rhizogonium Brid	RHIZOGONIACEAE
Rhizohypnella Fleisch	HYPNACEAE
Rhizohypnum (Hampe) Herz. = *Mittenothamnium*	HYPNACEAE
Rhizomnium (Broth.) T. Kop	MNIACEAE
Rhizonium Endl., orthogr. var. *Rhizogonium*	RHIZOGONIACEAE
Rhizopelma C. Müll. ≡ *Rhizogonium*	RHIZOGONIACEAE
Rhodobryum Hampe	BRYACEAE
Rhynchocarpidium P. de la Varde & Leroy = *Schimperella*	BRACHYTHECLACEAE
Rhyncho-hypnum (Hampe) Hampe = *Rhynchostegium*	BRACHYTHECLACEAE
Rhynchostegiella (B.S.G.) Limpr	BRACHYTHECLACEAE
Rhynchostegiopsis C. Müll	HOOKERIACEAE
Rhynchostegium B.S.G	BRACHYTHECIACEAE
Rhystophyllum Ehrh. ex Britt. ≡ *Neckera*	NECKERACEAE
Rhytidiadelphus (Limpr.) Warnst	RHYTIDIACEAE
Rhylidiopsis Broth	RHYTIDIACEAE
Rhyddium (Sull.) Kindb	RHYTIDIACEAE
Rigodiadelphus Dix	LESKEACEAE
Rigodiopsis Dix. & Thér. = *Okamuraea*	RHYTIDIACEAE
Rigodium Kunze ex Schwaegr	LEMBOPHYLLACEAE

Roellia Kindb	BRYACEAE (Mniaceae)
Rotteria Wijk, Marg. & Florsch., orthogr. var. Rottleria ≡ *Hyophila*	POTTIACEAE
Rottleria Brid., hom. illeg. ≡ *Hyophila*	POTTIACEAE
Rozea Besch	HYPNACEAE (Entodontaceae)
Rutenbergia Geh. & Hampe in Besch	RUTENBERGIACEAE
Saelania Lindb	DITRICHACEAE
Sainsburia Dix. = *Fissidens*	FISSIDENTACEAE
Sakuraia Broth. = *Eniodon*	ENTODONTACEAE
Sanioa Fleisch., orthogr. var. Sanionia = *Drepanocladus*	AMBLYSTEGIACEAE
Sanionia Loeske = *Drepanocladus*	AMBLYSTEGIACEAE
Saproma Brid., nom. illeg. = *Bruchia*	DICRANACEAE
Sarconeurum Bryhn	POTTIACEAE
Sasaokaea Broth. = *Drepanocladus*	AMBLYSTEGIACEAE
Sauloma (Hook. f. & Wils.) Mitt.	HOOKERIACEAE
Saviczia A. Abr. & I. Abr. = *Plagiothecium*	PLAGIOTHECIACEAE
Scabridens Bartr	LEUCODONTACEAE
Schimperella Thér	BRACHYTHECLACEAE
Schimperobryum Marg	HOOKERIACEAE
Schisthymenium Harv. in Saut., orthogr. var. Schizymenium = *Mielichhoferia*	BRYACEAE
Schistidium Brid	GRIMMIACEAE
Schistomitrium Dozy & Molk	DICRANACEAE (Leucobryaceae)
Schistophyllum Lindb., nom. illeg. = *Fissidens*	FISSIDENTACEAE
Schistostega Mohr	SCHISTOSTEGACEAE
Schistotega Dum., orthogr. var. *Schistostega*	SCHISTOSTEGACEAE
Schizhymenium C. Müll., orthogr. var. Schizymenium = *Mielichhoferia*	BRYACEAE
Schizodon Sw., nom. illeg. = *Schlotheimia*	ORTHOTRICHACEAE
Schizomitrium B.S.G	HOOKERIACEAE
Schizymenium Harv. in Hook. = *Mielichhoferia*	BRYACEAE
Schleitemia Dism., orthogr. var. *Schlotheimia*	ORTHOTRICHACEAE
Schliephackea C. Müll	DICRANACEAE
Schlotheimia Brid.	ORTHOTRICHACEAE
Schoenobryum Dozy & Molk	CRYPHAEACEAE
Schraderella C. Müll	SEMATOPHYLLACEAE
Schraderobryum Fleisch	SEMATOPHYLLACEAE
Schroeterella Herz	SEMATOPHYLLACEAE
Schweischkea C. Müll	FABRONIACEAE
Schwetschkeopsis Broth	FABRONIACEAE
Schyzimenium Harv., orthogr. var. Schizymenium = *Mielichhoferia*	BRYACEAE
Sciadocladus Lindb. ex Broth. = *Hypnodendron*	HYPNODENDRACEAE
Sciaromiopsis Broth	AMBLYSTEGIACEAE
Sciaromium (Mitt.) Mitt	AMBLYSTEGIACEAE
Sciuro-hypnum (Hampe) Hampe = *Brachythecium*	BRACHYTHECIACEAE
Sciuroleskea Hampe ex Broth	BRACHYTHECLACEAE
Sclerodontium Schwaegr	DICRANACEAE

Sclerohypnum Dix	FONTINALACEAE
Scleromnium Jur. = *Echinodium*	ECHINODIACEAE
Scleropodium B.S.G	BRACHYTHECIACEAE
Scopelophila (Mitt.) Lindb	POTTIACEAE
Scorpidium (Schimp.) Limpr	AMBLYSTEGIACEAE
Scorpiurium Schimp	BRACHYTHECIACEAE
Scorpiurus Le Jol., orthogr. var. *Scorpiurium*	BRACHYTHECIACEAE
Scoulera Hag., orthogr. var. *Scouleria*	GRIMMIACEAE
Scouleria Hook, in Drumm	GRIMMIACEAE
Scytalina Hag., nom. illeg. ≡ *Orthodicranum*	DICRANACEAE
Sebillea Biz	POTTIACEAE
Sekra Adans. ex Lindb., nom. illeg. ≡ *Cinclidotus*	POTTIACEAE
Seligera Schur., orthogr. var. *Seligeria*	SELIGERIACEAE
Seligeria B.S.G	SELIGERIACEAE
Sematophyllum Mitt	SEMATOPHYLLACEAE
Semibarbula Herz, ex Hilp	POTTIACEAE
Serpoleskea (Limpr.) Loeske ≡ *Platydictya*	AMBLYSTEGIACEAE
Serpotortella Dix	POTTIACEAE
Sharpiella Iwats. = *Henogiella*	HYPNACEAE
Sigmatella (C. Müll.) C. Müll., horn, illeg. = *Taxithelium*	SEMATOPHYLLACEAE
Simophyllum Lindb., nom. illeg. ≡ *Weissia*	POTTIACEAE
Simplicidens Herz. = *Fissidens*	FISSIDENTACEAE
Sinocalliergon Sak	AMBLYSTEGIACEAE
Skisophyllum Wijk, Marg. & Florsch., orthogr. var. Skitophyllum ≡ *Fissidens*	FISSIDENTACEAE
Skitophyllum B. Pyl., nom. illeg. ≡ *Fissidens.*	FISSIDENTACEAE
Skottsbergia Card	DITRICHACEAE
Skytophyllum Wittst., orthogr. var. Skitophyllum ≡ *Fissidens*	FISSIDENTACEAE
Solmsia Hampe, hom. illeg. = *Braunfelsia*	DICRANACEAE
Solmsiella C. Müll. = *Erpodium*	ERPODIACEAE
Sorapilla Spruce ex Mitt	SORAPILLACEAE
Sphaerangium Schimp., nom. illeg. ≡ *Astomum*	POTTIACEAE
Sphaerocephalus Neck, ex Lindb. ≡ *Aulacomnium*	AULACOMNIACEAE
Sphaerotheciella Fleisch	CRYPHAEACEAE
Sphaerothecium Hampe	DICRANACEAE
Sphagnum L	SPHAGNACEAE
Spiridens Nees	SPIRIDENTACEAE
Spiridentopsis Broth	PTEROBRYACEAE
Spirula Dozy & Molk. = *Cladopodanthus*	DICRANACEAE
Splachnobryum C. Müll.	POTTIACEAE (Splachnaceae)
Splachnum Hedw	SPLACHNACEAE
Sporledera Hampe (Ditrichaceae) = *Bruchia*	DICRANACEAE
Sprucea Hook. f. &Wils., nom. illeg. =*Holomitrium*	DICRANACEAE
Spruceella C. Müll., hom. illeg	POTTIACEAE
Squamidium (C. Müll.) Broth	METEORIACEAE
Stablera Hag., orthogr. var. Stableria = *Orthodontium*	BRYACEAE
Stableria Lindb. ex Braithw. = *Orthodontium*	BRYACEAE
Steereobryon G. L. Sm	POLYTRICHACEAE
Stegonia Vent	POTTIACEAE

Stenocarpidiopsis Fleisch. ex Broth	PLAGIOTHECIACEAE
Stenocarpidium C. Müll	PLAGIOTHECIACEAE
Stenodesmus (Mitt.) Jaeg	DALTONIACEAE (Hookeriaceae)
Stenodiclyon (Mitt.) Jaeg	HOOKERIACEAE
Stenodyction Fleisch., orthogr. var. *Stenodiclyon*	HOOKERIACEAE
Stenomitrium (Mitt.) Broth	ORTHOTRICHACEAE
Stenotheciopsis Fleisch	HYLOCOMIACEAE
Stephanodictyon Dix	POTTIACEAE
Stephanostoma (Mitt.) Kindb. = *Erpodium*	ERPODIACEAE
Steppomitra Vond. & Hadač	FUNARIACEAE
Stereodon Mitt., nom. illeg. ≡ *Hypnum*	HYPNACEAE
Stereodontopsis Williams	HYPNACEAE
Stereohypnum (Hampe) Fleisch. ≡ *Miltenothamnium*	HYPNACEAE
Stereophyllum Mitt	PLAGIOTHECIACEAE
Steyermarkiella Robins	DICRANACEAE
Stigmatella Reed & Robins., orthogr. var Sigmatella = *Taxithelium*	SEMATOPHYLLACEAE
Stirtonia R. Br. ter. = *Trematodon*	DICRANACEAE
Stokesiella (Kindb.) Robins	BRACHYTHECIACEAE
Streblopilum Angstr. = *Brachythecium*	BRACHYTHECIACEAE
Streblotrichum P. Beauv	POTTIACEAE
Streptocalypta C. Müll	POTTIACEAE
Streptocarpus Poech in Keil., hom. illeg. = *Encalypta*	ENCALYPTACEAE
Streptopogon Wils, in Mitt	POTTIACEAE
Streptotrichum Herz	POTTIACEAE
Striptopogon Hampe, orthogr. var. *Streptopogon*	POTTIACEAE
Stroemia I. Hag., hom. illeg. = *Orthotrichum*	ORTHOTRICHACEAE
Struckia C. Müll	SEMATOPHYLLACEAE
Stylostegium B.S.G	SELIGERIACEAE
Swartzia Brid., hom. illeg. = p.p. *Distichium*	DITRICHACEAE
Symblepharis Mont	DICRANACEAE
Symphyodon Mont	DALTONIACEAE (Symphyodontaceae)
Symphysodon Dozy & Molk	PTEROBRYACEAE
Symphysodoniella Fleisch	PTEROBRYACEAE
Sympoma Aust. = *Micromitrium*	EPHEMERACEAE
Synodontella Dix. & Then = *Entodon*	ENTODONTACEAE
Synodontia (Besch.) Broth	DICNEMONACEAE
Synthetodontium Card	BRYACEAE
Syntrichia Brid. = *Toriula*	POTTIACEAE
Syrigothecium Mitt., orthogr. var. *Syringothecium*	SEMATOPHYLLACEAE
Syringothecium Mitt	SEMATOPHYLLACEAE
Syrrhopodon Schwaegr	CALYMPERACEAE
Systegium Schimp., nom. illeg. ≡ *Astomum*	POTTIACEAE
Systilium Hook., orthogr. var. Systylium = *Tayloria*	SPLACHNACEAE
Systylium Homsch. = *Tayloria*	SPLACHNACEAE
Taiwanobryum Nog	PRIONODONTACEAE
Tamariscella (C. Müll.) C. Müll. ≡ *Thuidium*	THULDLACEAE
Taxicaulis (C. Müll.) C. Müll. = ?*Isopterygium*	HYPNACEAE

Taxiphyllum Fleisch	HYPNACEAE
Taxitheliella Dix	FABRONIACEAE
Taxithelium Spruce ex Mitt	SEMATOPHYLLACEAE
Tayloria Hook	SPLACHNACEAE
Teichodontium C. Müll. = *Macromitrium*	ORTHOTRICHACEAE
Teniolophora Reese	POTTIACEAE
Teretidens Williams = *Wilsoniella*	DICRANACEAE
Tetracladium (Mitt.) Broth., hom. illeg. ≡ *Bryonoguchia*	THUIDIACEAE
Tetracoscinodon R. Br. ter. = *Eucladium*	POTTIACEAE
Tetradontium Angstr., orthogr. var. Tetrodontium = *Tetraphis*	TETRAPHIDACEAE
Tetraphidopsis Broth. & Dix	PTYCHOMNIACEAE
Tetraphis Hedw	TETRAPHIDACEAE
Tetraplodon B.S.G	SPLACHNACEAE
Tetrapterum Hampe ex Jaeg	POTTIACEAE
Tetrastichium (Mitt.) Card	HOOKERIACEAE
Tetrastichium (Broth.) Fleisch., hom. illeg. ≡ *Bryonoguchia*	THUIDIACEAE
Tetrodontium Schwaegr. = *Tetraphis*	TETRAPHIDACEAE
Thamniella Besch. = *Camptochaete*	LEMBOPHYLLACEAE
Thamniopsis (Mitt.) Fleisch	HOOKERIACEAE
Thamnium B.S.G., hom. illeg. ≡ *Thamnobryum*	NECKERACEAE
Thamnobryum Nieuwl	NECKERACEAE
Thedenia B.S.G., hom. illeg. = *Pylaisia*	HYPNACEAE
Thelia Sull, in Sull. & Lesq	THELIACEAE
Theriotia Card	BUXBAUMIACEAE (Diphysciaceae)
Thesanomitrium Arnott, orthogr. var. Thysanomitrion = *Campylopus*	DICRANACEAE
Thiemea C. Müll	DICRANACEAE
Thuidiopsis (Broth.) Fleisch. = *Thuidium*	THUIDIACEAE
Thuidium B.S.G	THUIDIACEAE
Thujidium C. Jens., orthogr. var. *Thuidium*	THUIDIACEAE
Thuyidium Lindb., orthogr. var. *Thuidium*	THUIDIACEAE
Thydium Mann, orthogr. var. *Thuidium*	THUIDIACEAE
Thyidium Lindb., orthogr. var. *Thuidium*	THUIDIACEAE
Thyridiutn Mitt., hom. illeg. ≡ *Mitthyridium*	CALYMPERACEAE
Thysanomitrion Schwaegr. = *Camplyopus*	DICRANACEAE
Thysanomitriopsis C. Müll. = *Campylopus*	DICRANACEAE
Thysanomitrium Schwaegr., orthogr. var. Thysanomitrion = *Campylopus*	DICRANACEAE
Tichodontium Hag., orthogr. var. Teichodontium = *Macromitrium*	ORTHOTRICHACEAE
Timmea Schur, orthogr. var. *Timmia*	TIMMIACEAE
Timmia Hedw., nom. cons	TIMMIACEAE
Timmiella (De Not.) Limpr	POTTIACEAE
Tisserantiella P. de la Varde	POTTIACEAE
Tomenthypnum Loeske, orthogr. var. *Tomentypnum*	BRACHYTHECIACEAE
Tomentypnum Loeske	BRACHYTHECIACEAE
Tortella (Lindb.) Limpr., nom. cons.	POTTIACEAE
Tortula Hedw., nom. cons	POTTIACEAE
Trachybryum (Broth.) Schof	BRACHYTHECIACEAE

Trachycarpidium Broth. .... POTTIACEAE
Trachycystis Lindb .... MNIACEAE
Trachyloma Brid .... PTEROBRYACEAE
Trachymitrium Brid. = *Syrrhopodon* .... CALYMPERACEAE
Trachynotus (C. Müll.) Card, in Fleisch. =
Exodictyon.................... CALYMPERACEAE
Trachyphyllum Gepp in Hiern .... ENTODONTACEAE
Trachypodopsis Fleisch .... TRACHYPODACEAE
Trachypus Reinw. & Homsch .... TRACHYPODACEAE
Trachythecium Fleisch .... HYPNACEAE
Tremathodon P. Beauv., orthogr. var. *Trematodon* .... DICRANACEAE
Trematodon Michx .... DICRANACEAE
Trentepohlia A. Roth, nom. rejic. = *Pohlia* .... BRYACEAE
Trichodon Schimp .... DITRICHACEAE
Tricholepis Kindb., nom. rejic. = *Papillaria* .... METEORIACEAE
Trichopilum Schimp. ex C. Müll. = *Pogonatum*.................... POLYTRICHACEAE
Trichosteleum Mitt .... SEMATOPHYLLACEAE
Trichostomopsis Card.................... POTTIACEAE
Trichostomum Bruch, nom. cons. .... POTTIACEAE
Trichostomum Hedw., nom. rejic .... POTTIACEAE
Tridontium Hook. f. in Hook .... POTTIACEAE
Trigonodictyon Dix. & P. de la Varde .... ORTHOTRICHACEAE
Triodontium I. Hag., orthogr. var. *Tridontium* .... POTTIACEAE
Triplocoma B. Pyl., nom. illeg. ≡ *Dawsonia* .... POLYTRICHACEAE
Tripterocladium (C. Müll.) Jaeg.................... LEMBOPHYLLACEAE
Triquetrella C. Müll .... POTTIACEAE
Trismegistia (C. Müll.) C. Müll.................... SEMATOPHYLLACEAE
Tristichella Dix .... SEMATOPHYLLACEAE
Tristichiopsis C. Müll. = *Tristichium* .... DITRICHACEAE
Tristichium C. Müll .... DITRICHACEAE
Trochobryum Breidl. & Beck .... SELIGERIACEAE
Trolliella Herz.................... SEMATOPHYLLACEAE
Tuerckheimia Broth.................... POTTIACEAE
Tutigaea Ando =*Hondaella* .... HYPNACEAE

Ulea C. Müll., hom. illeg.................... POTTIACEAE
Uleobryum Broth.................... POTTIACEAE
Uleopsis Thér .... POTTIACEAE
Ulota Mohr.................... ORTHOTRICHACEAE
Usmania Laz. = *Grimmia* .... GRIMMIACEAE

Valdiviella Thér. & Herz, in Herz .... DICRANACEAE
Vebera Kindb., orthogr. var. Webera Hedw. =
Pohlia .... BRYACEAE
Venturiella C. Müll .... ERPODIACEAE
Verrucidens Card.................... SELICERIACEAE
Vesicularia (C. Müll.) C. Müll .... HYPNACEAE
Vesiculariopsis Broth .... HOOKERIACEAE (Leucomiaceae)
Viridivellus Stone .... VIRIDIVELLERACEAE
Voitia Homsch.................... SPLACHNACEAE

Walkeria Hornsch., nom. rejic. = *Leucoloma*	DICRANACEAE
Warburgiella C. Müll, in Broth	SEMATOPHYLLACEAE
Wardia Harv. & Hook	WARDIACEAE (Fontinalaceae)
Warnstorfia Loeske = *Drepanocladus*	AMBLYSTEGIACEAE
Webera Hedw., hom. illeg. = *Pohlia*	BRYACEAE
Webera Ehrh. ex Lindb., hom. illeg. ≡ *Diphyscium*	BUXBAUMIACEAE
Weberus Krause, nom. illeg. = *Pohlia*	BRYACEAE
Weisia Spreng., orthogr. var. *Weissia* Hedw	POTTIACEAE
Weisiodon Schimp., nom. rejic. = *Gyroweisia*	POTTIACEAE
Weisiopsis Broth	POTTIACEAE
Weissia Hedw	POTTIACEAE
Weissia Ehrh. ex Gaertn., hom. illeg. = *Vlota*	ORTHOTRICHACEAE
Werneriobryum Herz	DICNEMONACEAE
Weymouthia Broth	METEORIACEAE
Wijkia Crum	SEMATOPHYLLACEAE
Wildia C. Müll. & Broth	ERPODIACEAE
Willia C. Müll, in Neum	POTTIACEAE
Williamsia Broth., hom. illeg. = *Leptodontium*	POTTIACEAE
Williamsiella Britt. = *Leptodontium*	POTTIACEAE
Wilsoniella C. Müll	DICRANACEAE
Wollnya Herz. = *Leptobryum*	BRYACEAE
Zelometeorium Manuel	METEORIACEAE
Zieria Schimp., hom. illeg. = *Plagiobryum*	BRYACEAE
Zygodon Hook. & Tayl	ORTHOTRICHACEAE
Zygotrichia Brid. = *Desmatodon*	POTTIACEAE

1 SPHAGNACEAE Dum.
2 ANDREAEACEAE Dum.
2 A ANDREAEOBRYACEAE Stcere & B. Murr.
3 FISSIDENTACEAE Schimp.
3 A NANOBRYACEAE Schultze-Mot.
4 ARCHIDIACEAE Schimp.
5 DITRICHACEAELimpr., nom. cons. prop.
5 A VIRIDIVELLERACEAE Stone
6 BRYOXIPHIACEAEBescb., nom. cons.
7 SELIGERIACEAE Schimp.
8 DICRANACEAE Schimp.
9 DICNEMONACEAEBroth.
10 PLEUROPHASCACEAEBroth.
11 Leucobryaceae Schimp.
12 CALYMPERACEAE Kindb.
13 ENCALYPTACEAE Schimp.
14 POTTIACEAESchimp., nom. cons. prop.
14A BRYOBARTRAMIACEAE Sainsb.
15 GRIMMIACEAE Arnott
16 GIGASPERMACEAEBroth.
17 DISCELIACEAE Schimp.
18 EPHEMERACEAE Schimp.
19 FUNARIACEAE Schwaegr.
19A PsEUDODITRrCHACEAE Steere & Iwats.[1]
20 OEDIPODIACEAE Schimp.
21 SPLACHNACEAE Grev.& Arnott
22 SCHISTOSTEGACEAE Schimp.
23 TETRAPHIDACEAE Schimp.
24 BRYACEAE Schwaegr.
25 Leptostomataceae Schwaegr.
26 MNIACEAE Schwaegr.
27 PHYLLODREPANIACEAE Crosby
28 EUSTICHIACEAEBroth., nom. cons.
29 SoBAPILLACEAE Fleisch.
30 MITTEN IACEAEBroth.
31 CALOMNIACEAE Kindb.
32 RHIZOGONIACEAEBroth.
33 HYPNODENDRACEAEBroth.
34 AULACOMNIACEAE Schimp.
35 MEESIACEAE Schimp.
36 CATOSCOPIACEAEBroth.
37 BARTRAMIACEAE Schwaegr.
38 SPIRIDENTACEAE Kindb.
39 TIMMIACEAE Schimp.
40 ERPODIACEAEBroth.
40A RHACHITHECIACEAE Robins.
40B MICROTHECIELLACEAE Mill.& Harr.41
PTYCHOMITRIACEAE Schimp.
42 ORTHOTRJCHACEAE Arnott 42A
RHABDOWEISIACEAE Limpr. 43
HELICOPHYLLACEAEBroth.
44 RACOPILACEAE Kindb.
45 FONTINALACKAE Schimp.
4SA WARDIACEAE Welch[2]
45B HYDROPOGONACEAE Welch[2]
46 CLIMACIACEAE Kindb.
46A PLEUROZIOPSACEAE Irel.[3]
47 HEDWIGIACEAE Schimp.
48 CRYPHAEACEAE Schimp.
49 LEUCODONTACEAE Schimp.
SO CYRTOPODACEAE Fleisch.
51 PTYCHOMNIACEAE Fleisch.
52 LEPYRODONTACEAEBroth.
53 PRIONODONTACEAEBroth.
54 RUTENBERGIACEAE (Broth.)Fleisch.
55 TBACHYPODACEAE Fleisch.
56 MYURIACEAE Fleisch.
57 PTEROBRYACEAE Kindb.
58 METEORIACEAE Kindb.
59 PHYLLOGONIACEAE Kindb.
60 NECKERACEAE Schimp.
60A Thamniaceae Monk., nom. illeg.
60B DALTONIACEAE Schimp.[4]
61 LEMBOPHYLLACEAEBroth.
62 ECHINODIACKAEBroth.
63 Ephemeropsaceae Schultze-Mot.
64 Pdotrichaceae Kindb.
65 HOOKERIACEAE Schimp.
65A Distichophyllaceae Mill.
66 Symphyodontaceae Fleisch.
67 Leucomiaceae Broth.
68 Hypopterygiaceae Mitt
68A Cyathophoraceae Kindb.
69 THELIACEAE (Broth.)Fleiscb.
70 FABRONIACEAE Schimp.
71 LESKEACEAE Schimp.
71A REGMATODONTACEAE Broth.
72 THUIDIACEAE Schimp.
72A Cratoneuraceae Mönk.
73 AMBLYSTEGIACEAE (Broth.)Fleisch.
74 BRACHYTHECIACEAE Schimp.
75 ENTODONTACEAEKindb., nom. cons. prop.
76 PLAGIOTHECIACEAE (Broth.)Fleisch.
77 SEMATOPHYLLACEAEBroth., nom. cons. prop.
78 HYPNACEAE Schimp.
79 RHYTIDIACEAEBroth.
80 HYLOCOMIACEAE (Broth.)Fleisch.
81 BUXBAUMIACEAE Schwaegr.
82 Diphysciaceae Fleiscb.
83 POLYTRICHACEAE Schwaegr.
84 Dawsoniaceae Broth.

---

[1] Steere and Iwatsuki (1974) did not indicate the relationships of this family when originally proposing it We have placed it near Funariaceae on the basis of their statement that the peristome is "perhaps at the evolutionary level of the Funariaceae"

[2] Welch (1943) did not indicate the relationships of these families when she segregated them from Fontinalaceae, and we have placed them here for convenience. Robinson (1971) placed the former near Hedwigiaceae and the latter at the end of his classification.

[3] Ireland (1968) did not indicate the relationships of this family when he segregated it from Climaciaceae. Robinson (1971) placed it after Climaciaceae.

[4] Discussed by Crosby (1974), but their relationships were not indicated; we place it here on basis of the peristome.

Amblyodontaceae Schimp., Syn. Muse. Eur., ed. 2. CII, 495. 1876, "Amblyodonteae." Type: *Amblyodon* B.S.G., nom. cons. = MEESIACEAE.

AMBLYSTEGIACEAE G. Roth, Hedwigia 38: (6). 1899. Type: *Amblystegium* B.S.G.

ANDREAEACEAE Dum., Anal. Fam. Pl. 68. 1829. Type: *Andreaea* Hedw.

ANDREAEOBRYACEAE Steere in Steere & B. Murr., Phytologia 33:410. 1976. Type: *Andreaeobryum* Steere & B. Murr.

Anoectangiaceae Schimp., Coroll. Bryol. Eur. 11. 1855 [1856], Type: *Anoectan- gium* Schwaegr., nom. cons. = POTTIACEAE.

Anomodontaceae Kindb., Gen. Eur. Northamer. Bry. 6. 1897. Type: *Anomodon* Hook. & Tayl. = THUIDIACEAE.

Aongstroemiaceae De Not., Epil. Briol. Ital. 30. 1869, "Angstroemiacei." Type: *Aongstoremia* B.S.G. = DICRANACEAE.

ARCHIDIACEAE Schimp., Coroll. Bryol. Eur. 5. 1855 [1856]. Type: *Archidium* Brid.

Archifissidentaceae Dix. & P. de la Varde, Arch. Bot. Mém. 3: 23. 1927, nom. inval. Based on: *Nanobryum* Dix. ≡ NANOBRYACEAE.

Astomataceae Schimp., Coroll. Bryol. Eur. 7. 1855 [1856], "Astomaceae." Type: *Astomum* Hampe = POTTIACEAE.

AULACOMNIACEAE Schimp., Syn. Musc. Eur. 411. 1860, "Aulacomnieae." Type: *Aulacomnium* Schwaegr., nom. cons.

BARTRAMIACEAE Schwaegr. in Willd., Sp. Pl., ed. 4. 5(2): 90. 1830, "Bartram- iae." Type: *Bartramia* Hedw., nom. cons.

Blindiaceae Schimp., Syn. Musc. Eur. 117. 1860, "Blindieae." Type: *Blindia* B.S.G. = SELIGERIACEAE.

Brachyodontaceae Schimp., Syn. Musc. Eur. 116. 1860, nom. illeg., "Brachy- odonteae." Type: *Brachyodus* Fürnr. ex Nees & Homsch., nom. illeg. = SELIGERIACEAE.

BRACHYTHECIACEAE Schimp., Syn. Muse. Eur., ed. 2. XCV *sic* [CXV], 637. 1876,"Brachythecieae." Type: *Brachythecium* B.S.G.

Bruchiaceae Schimp., Coroll. Bryol. Eur. 6. 1855 [1856]. Type: *Bruchia* Schwaegr. = DICRANACEAE.

BRYACEAE Schwaegr. in Willd., Sp. Pl., ed. 4. 5(2): 47. 1830, "Bryi." Type: *Bryum* Hedw.

BRYOBARTRAMIACEAE Sainsb., Bryologist 51: 9. 1948. Type: *Bryobartramia* Sainsb.

BRYOXIPHIACEAE Besch., Jour. Bot. (Morot) 6: 183. 1892, nom. cons., "Bryoxiphieae." Type: *Bryoxiphium* Mitt., nom. cons.

BUXBAUMIACEAE Schwaegr. in Willd., Sp. Pl., ed. 4. 5(2): 23. 1830, "Bux- baumiae." Type: *Buxbaumia* Hedw.

CALOMNIACEAE Kindb., Bot. Centralbl. 77: 394. 1899. Type: *Calomnion* Hook, f. & Wils..

CALYMPERACEAE Kindb., Gen. Eur. Northamer. Bry. 11.1897. Type: *Calymperes* Sw. in Web.

Camptotheciaceae Schimp., Syn. Muse. Eur., ed. 2. XCV, 635. 1876, "Camp- tothecieae." Type: *Camptothecium* B.S.G. = BRACHYTHECIACEAE.

Campylosteliaceae De Not., Epil. Briol. Ital. 31. 1869, "Campylosteliacei." Type: *Campylostelium* B.S.G. =PTYCHOMITRIACEAE.

CATOSCOPIACEAE Boul. ex Broth., Nat. Pflanzenfam. I(3):629. 1904. Type: *Cato- scopium* Brid.

Ceratodontaceae Schimp., Syn. Musc. Eur. 138. 1860, nom. rejic. prop., "Cerato- donteae." Type: *Ceratodon* Brid. = DITRICHACEAE. [This name antedates the more frequently used Ditrichaceae Limpr. and was published simultaneously with Distichiaceae Schimp.]

Cinclidiaceae Kindb., Gen. Eur. Northamer. Bry. 12. 1897. Type:*Cinclidium* Sw. in Schrad. = MNIACEAE.

Cinclidotaceae Schimp., Syn. Musc. Eur. 193. 1860, "Cinclidoteae." Type: *Cinclidotus* P. Beauv. = POTTIACEAE.

CUMACIACEAE Kindb., Gen. Eur. Northamer. Bry. 7. 1897. Type: *Climacium* Web. & Mohr. [Although Kindberg included the type of Isotheciaceae, *Isothecium* Brid., that family name is illegitimate, and Climaciaceae is legitimate and correct.]

Cratoneuraceae Mönk. in Rabenh., Kryptogamen-Fl. 4(Ergänzungsb.): 704. 1927. Type: *Cratoneuron* (Sull.) Spruce = AMBLYSTEGIACEAE.

CRYPHAEACEAE Schimp., Coroll. Bryol. Eur. 97. 1855 [1856]. Type: *Cryphaea* Mohr in Web.

Cryptotheciaceae Berk., Handb. Brit. Mosses 65. 1863, nom. inval., "Cryp- tothecii." Type: *Cryphaea* Mohr in Web. ≡ CRYPHAEACEAE. [Although the name *Cryptothecium,* for a fossil moss, antedates the publication of this name, it appears to be typified by *Cryphaea,* since this is the only genus Berkeley mentioned.]

Cyathophoraceae Kindb., Bot. Centralbl. 76: 85. 1898. Type: *Cyathophorum* P. Beauv. = DALTONIACEAE.

Cylindrotheciaceae Schimp., Syn. Musc. Eur. 510. 1860, nom. illeg., "Cylin- drothecieae." Type: *Cylindrothecium* B.S.G., nom. illeg. = ENTODONTA- CEAE.

CYRTOPODACEAE Fleiscb., Musci Fl. Buitenzorg 3: 655. 1908. Type: *Cyrtopus* (Brid,) Hook. f.

DALTONIACEAE Schimp., Syn. Musc. Eur. 478. 1860, "Daltonieae." Type: *Dal- tonia* Hook. & Tayl., nom. cons. prop.

Dawsoniaceae Broth., Nat. Pflanzenfam. I(3): 698. 1905. Type: *Dawsonia* R. Br. = POLYTRICHACEAE.

Dendroideaceae G. Roth, Hedwigia 38: (7). 1899, nom. inval. Based on *DendroHypnum* Hampe, nom. inval. = HYPNODENDRACEAE.

Dichelymaceae Schimp., Syn. Musc. Eur. 458. 1860, "Dichelymeae." Type: *Dichelyma* Myr. = FONTINALACEAE.

DICNEMONACEAE Broth., Nat. Pflanzenfam., ed. 2. 10: 214. 1924. Type: *Dicne- mon* Schwaegr.

DICRANACEAE Schimp., Coroll. Bryol. Eur. 11. 1855 [1856], Type: *Dicranum* Hedw.

Diphysciaceae Fleisch. in Engl., Syllabus Pflanzenfam., ed. 8. 88. 1919. Type: *Diphyscium* Mohr = BUXBAUMIACEAE.

DISCELIACEAE Schimp., Coroll. Bryol. Eur. 59. 1855 [1856]. Type: *Discelium* Brid.

Distichiaceae Schimp., Syn. Musc. Eur. 135. 1860, nom. rejic. prop. "Distichieae." Type: *Distichium* B.S.G. = DITRICHACEAE. [This name antedates the more frequently used Ditrichaceae Limpr. and was published simultaneously with Ceratodontaceae Schimp.]

Distichophyllaceae Mill., Phytologia 21:249. 1971. Type:*Distichophyllum* Dozy& Molk. = HOOKERIACEAE.

DITRICHACEAE Limpr. in Rabenh., Kryptogamen-Fl., ed. 2. 4: 482. 1887, nom. cons. prop. Type: *Ditrichum* Hampe. [When published, this commonly used name included the types of Ceratodontaceae Schimp. and Distichiaceae Schimp., both of which date from 1860.]

Drepanophyllaceae Mont. ex Berk., Handb. Brit. Mosses 156. 1863, nom. illeg., "Drepanophyllei." Type: *Drepanophyllum* Schwaegr., hom, illeg. ≡ PHYL- LODREPANIACEAE. [Although Berkeley treated only *Fissidens,* he attributed the name to Montagne, whose use of it was invalid but who included *Drepanophyllum*.]

ECHINODIACEAE Broth., Nat. Pflanzenfam. I(3): 1216. 1909. Type: *Echinodium* Jur.

ENCALYPTACEAE Schimp., Coroll. Bryol. Eur. 38. 1855 [1856]. Type: *Encalypta* Hedw.

Endotrichaceae Kindb., Gen. Eur. Northamer. Bry. 6. 1897, nom. illeg. Type: *Endotrichum* Dozy & Molk., hom. illeg. = PTEROBRYACEAE.

ENTODONTACEAE Kindb., Gen. Eur. Northamer. Bry. 7. 1897, nom. cons. prop.
Type: *Entodon* C. Müll.

EPHEMERACEAE Schimp., Coroll. Bryol. Eur. 3. 1855 [1856], "Ephemereae." Type: *Ephemerum* Hampe, nom. cons.

Ephemerellaceae Schimp., Syn. Musc. Eur., ed. 2. LXXXIV, 9. 1876, "Ephemerelleae." Type: *Ephemerella* C. Müll. = EPHEMERACEAE.

Ephemeropsaceae Schultze-Mot , Taxon 19: 251. 1970, "Ephemeropsidaceae."
Type: *Ephemeropsis* Goeb. = HOOKERIACEAE. [Published as a replacement for Nemataceae, nom. inval.]

ERPODIACEAE Broth., Nat. Pflanzenfam. I(3): 706. 1905. Type: *Erpodium* (Brid.) Brid.

Erpodiopsaceae C. Müll., Flora 73:470. 1890, "Erpodiopsideae." Type: *Erpodiop- sis* C. Müll. = POTTIACEAE.

EUSTICHIACEAE Broth., Nat Pflanzenfam., ed. 2. 10: 420. 1924, nom. cons. Type: *Eustichia* (Brid.) Brid.

Eustichiaceae Schimp., Coroll. Bryol. Eur. 37. 1855 [1856], nom. rejic. Type: *Eustichium* B.S.G., nom. rejic. = BRYOXIPHIACEAE.

FABRONIACEAE Schimp., Coroll. Bryol. Eur. 102. 1855 [1856]. Type; *Fabronia* Raddi.

FISSIDENTACEAE Schimp., Coroll. Bryol. Eur. 20. 1855 [1856]. Type: *Fissidens* Hedw.

FONTINALACEAE Schimp., Coroll. Bryol. Eur. 96. 1855 [1856]. Type: *Fontinalis* Hedw.

FUNARIACEAE Schwaegr. in Willd., Sp. Pl., ed. 4. 5(2): 43. 1830, "Funariae." Type: *Funaria* Hedw.

Georgiaceae Kindb., Gen. Eur. Northamer. Bry. 9. 1897, nom. illeg. Type: *Georgia* C. Müll., nom. illeg. ≡ TETRAPHIDACEAE.

GIGASPERMACEAE Lindb., övers. Förh. Kongl. Svenska Vetensk.-Akad. 1864: 592. 1865. Type: *Gigaspermum* Lindb.

GRIMMIACEAE Arnott, Disp. Méth. Mousses 19. 1825, "Grimmoideae." Type: *Grimmia* Hedw.

Habrodontaceae Schimp., Syn. Musc. Eur. 504. 1860, "Habrodonteae." Type: *Habrodon* Schimp. = FABRONIACEAE.

Harrisoniaceae C. Müll., Oesterr. Bot. Zeit. 47: 419. 1897, nom. illeg. Type: *Harris onia* Spreng., hom. illeg. = HEDWIGIACEAE.

HEDWIGIACEAE Schimp., Coroll. Bryol. Eur. 52. 1855 [1856]. Type:*Hedwigia* P. Beauv., nom. cons.

HELICOPHYLLACEAE Broth., Nat. Pflanzenfam. I(3): 973. 1907. Type:*Helicophyl lum* Brid.

Homaliaceae Schimp., Syn. Musc. Eur. CXLIV. 1860, "Homalieae." Type: *Homalia* (Brid.) B.S.G. = NECKERACEAE.

HOOKERIACEAE Schimp., Coroll. Bryol. Eur. 101. 1855 [1856]. Type:*Hookeria* J. E. Sm., nom. cons.

HYDROPOCONACEAE Welch, Bryologist 46: 31. 1943. Type: *Hydropogon* Brid.

HYLOCOMIACEAE (Broth.) Fleisch., Nova Guinea 12: 125. 1914.

Hypnaceae subfam. Hylocomioïdeae Broth., Nat. Pflanzenfam. I(3): 1044. 1908, "Hylocomieae." Type: *Hylocomium* B.S.G.

Hyophilaceae C. Müll., Linnaea 20: 68. 1847, nom. rejic. prop. Type: *Hyophila* Brid. = POTTIACEAE. [This name antedates the widely used Pottiaceae Schimp.]

HYPNACEAE Schimp., Coroll. Bryol. Eur. 113. 1855 [1856]. Type: *Hypnum* Hedw., nom. cons.

HYPNODENDRACEAE Broth., Nat. Pflanzenfam. I(3): 1166. 1909. Type: *Hypnodendron* (C. Müll.) Lindb. ex Mitt.

Hypno-leskeaceae Schimp., Coroll. Bryol. Eur. 111. 1855 [1856], nom. inval. p.p. LESKEACEAE& p.p. THUIDIACEAE. [This family name is not based on a generic name.]

Hypopterygiaceae Mitt., Jour. Proc. Linn. Soc., Suppl. Bot. 1: 147. 1859. Type: *Hypopterygium* Brid. = HOOKERIACEAE.

Isotheciaceae Berk., Handb. Brit. Mosses 139. 1863, nom. illeg., "Isothecii." Based on: *hothecium* Brid. [The types of Leskeaceae Schimp. and Leucodontaceae Schimp., both of which antedate this name, were included by Berkeley.]

LEMBOPHYLLACEAE Broth., Nat. Pflanzenfam. I(3): 863. 1906. Type:*Lembophyl- lum* Lindb.

Leptodontaceae Schimp., Coroll. Bryol. Eur. 98. 1855 [1856], Type: *Leptodon* Mohr = NECKERACEAE.

Leptostomataceae Schwaegr. in Willd., Sp. Pl, ed. 4. 5(2): 85. 1830, "Leptostomi." Type: *Leptostomum* R. Br., nom. cons. = BRYACEAE.

Leptotrichaceae Schimp., Syn. Muse. Eur. XCI, 138. 1876, nom. illeg. Type: *Leptotrichum* Hampe ex C. Müll., nom. illeg. = DITRICHACEAE.

LEPYRODONTACEAE Broth., Nat. Pflanzenfam. 1(3): 771. 1906. Type: *Lepyrodon* Hampe.

LESKEACEAE Schimp., Coroll. Bryol. Eur. 109. 1855 [1856]. Type: *Leskea* Hedw.
Leucobryaceae Schimp., Coroll. Bryol. Eur. 19. 1855 [1856]. Type: *Leucobryum* Hampe = DICRANACEAE.

LEUCODONTACEAE Schimp., Coroll. Bryol. Eur. 108. 1855 [1856]. Type: *Leucodon* Schwaegr.

Leucomiaceae Broth., Nat. Pflanzenfam. 1(3): 1095. 1908. Type: *Leucomium* Mitt. = HYPNACEAE.

Leucophanaceae Fleisch., Musci Fl. Buitenzorg 1: 166. 1904. Type: *Leucophanes* Brid. = CALYMPERACEAE.

Lopidiaceae Rodw., Tasmanian Bryoph. 1:10. 1914, nom. illeg., "Lophidiaceae." Based on: *Lopidium* Hook. f. & Wils. [The types of Racopilaceae Kindb. and Hypopterygiaceae Mitt., both of which antedate this name, were included by Rodway.]

Macrohymeniaceae Kiaer, Forhandl. Vidensk.-Selsk. Christiana 1882(24); 6. 1883, nom. rejic. prop., "Macrohymenieae."Type: *Macrohymenium* C. Müll. = SEMATOPHYI.LACEAE. [*Macrohymenum* C. Müll (1847) is typified by *M. rufum* (Reinw. & Hornsch.)C. Müll., since the genus was monotypic when published. *Macrohymenium rufum* is generally treated as a species of *Acroporium* Mitt. (1868), but *Macrohymenium* has priority over *Acroporium.* Since *Acroporium* and *Sematophyllum* are placed in the same family, Macrohymeniaceae has priority over the better known Sematophyllaceae (1908).]

MEESIACEAE Schimp., Coroll. Bryol. Eur. 82. 1855 [1856]. Type: *Meesia* Hedw., nom. cons.

METEORIACEAE Kindb., Gen. Eur. Northamer. Bry. 7. 1897. Type: *Meteorium* (Brid.) Dozy & Molk.

Microthamniaceae Kindb., Bot. Centralbl. 77: 386. 1899, nom illeg. Type: *Microthamnium* Mitt., hom. illeg. = HYPNACEAE.

MICROTHECIELLACEAE Mill. & Harr., Jour. Bryol. 9: 523. 1977. Type: *Micro- theciella* Dix.

Mielichhoferiaceae Schimp., Co roll. Bryol. Eur. 62. 1855 [1856]. Type: *Mielich- hoferia* Nees & Hornsch. = BRYACEAE.

MITTENIACEAE Broth., Nat. Pflanzenfam. 1(3): 532. 1903. Type:*Mittenia* Lindb.

MNIACEAE Schwaegr. in Willd., Sp. Pl., ed. 4. 5(2): 25. 1830, "Mnii." Type: *Mnium* Hedw., nom. cons.

Mniadelphaceae Kindb., Bot. Centralbl. 76: 86. 1898, nom. illeg. Type:*Mniadel- phus* C. Müll., nom. illeg. = HOOKERIACEAE.

Myriniaceae Schimp., Syn. Musc. Eur., ed. 2. CXI, 589. 1876, "Myrinieae." Type: *Myrinia* Schimp., nom. cons. = FABRONIACEAE.

MYURIACEAE Fleisch., Musci Fl. Buitenzorg 4: 1663. 1923. Type: *Myurium* Schimp.

NANOBRYACEAE Schultze-Mot., Willdenowia 5: 386. 1969. Type: *Nanobryum* Dix. [Published as a replacement for Archifissidentaceae, nom. inval.]

NECKERACEAE Schimp., Coroll. Bryol. Eur. 99. 1855 [1856]. Type: *Neckera* Hedw., nom. cons.

Nemataceae Broth., Nat. Pflanzenfam. 1(3): 916. 1907, nom. inval., "Nematoceae." Type: *Ephemeropsis* Goeb. = HOOKERIACEAE. [Ephemeropsaceae has been published to replace Nemataceae.]

OEDIPODIACEAE Schimp., Syn. Musc. Eur., ed. 2. XCVIII, 354, "Oedipodieae." Type: *Oedipodium* Schwaegr.

Oreadaceae Berk., Handb. Brit. Mosses. 168. 1863, "Oradei." Type: *Oreas* Brid. = DICRANACEAE.

Orthotheciaceae Schimp., Coroll. Bryol. Eur. 103. 1855 [1856], nom. rejic. prop. Type: *Orthothecium* B.S.G., nom. cons. = ENTODONTACEAE. [This name antedates the more frequently used Entodontaceae.]

ORTHOTRICHACEAE Arnott, Disp. Méth. Mousses 13. 1825, "Orthotrichoideae." Type: *Orthotrichum* Hedw.

Paludellaceae De Not., Epil. Briol. Ital. 486. 1869, "Paludellacei." Type: *Palu- della* Ehrh. ex Brid. = MEESIACEAE.

Papillariaceae Kindb., Bot. Centralbl. 77: 386. 1899. Type: *Papillaria* (C. Müll.) C. Müll., nom. cons. = METEORIACEAE.

Phascaceae Schimp., Coroll. Bryol. Eur. 4. 1855 [1856]. Type: *Phascum* Hedw. = POTTIACEAE. [This name was published simultaneously with the generally used Pottiaceae.]

PHYLLODREPANIACEAE Crosby, Rev. Bryol. Lichénol. 37: 347. 1970 [1971]. Type: *Phyllodrepanium* Crosby. [Published as a replacement for Drepanophyl- laceae, nom. illeg.]

PHYLLOGONIACEAE Kindb., Bot. Centralbl. 76: 85. 1898. Type: *Phyllogonium* Brid.

Physcomitrellaceae Schimp., Syn. Musc. Eur., ed. 2. LXXXIV, 7. 1876, "Physcomitrelleae." Type: *Physcomitrella* B.S.G. = FUNARIACEAE.

Physcbmitriaceae Schimp., Syn. Musc. Eur. 312. 1860, nom. illeg., "Physco- mitrieae." Based on: *Physcamitrium* (Brid.) Fümr. = FUNARIACEAE. [The type of the earlier Funariaceae was included by Schimper when he published this name.]

Pilotrichaceae Kindb., Bot. Centralbl. 77: 54. 1899, nom. illeg. Type: *Pilotrichum* Brid., nom. illeg. = DALTONIACEAE.

PLAGIOTHECIACEAE (Broth.) Fleisch., Nova Guinea 8: 748. 1912.

Hypnaceae subfam. Plagiothecioïdeae Broth., Nat. Pflanzenfam. 1(3): 1078. 1908, "Plagiothecieae." Type: *Plagiothecium* B.S.G.

Pleuridiaceae Schimp., Coroll. Bryol. Eur. 6. 1855 [1856], nom. illeg. Type: *Pleuridium* Brid., nom. rejic. = ARCHIDIACEAE

Pleurobryaceae Schimp., Syn. Musc. Eur. 325. 1860, nom. inval. = BRYACEAE. [Schimper mentioned only *Mielichhqferia* when he published this name, but said that there were other exotic genera in the family. However, no generic name exists on which to base this family name.]

PLEUROPHASCACEAE Broth., Nat. Pflanzenfam. I(3): 774. 1906. Type: *Pleuro- phascum* Lindb.

PLEUROZIOPSACEAE Irel., Jour. Hattori Bot. Lab. 31: 62. 1968, "Pleuroziop- sidaceae." Type: *Pleuroziopsis* Kindb. ex Britt.

POLYTRICHACEAE Schwaegr. in Willd., Sp. Pl., ed. 4. 5(2):. 1. 1830, "Polytrichii." Type: *Polytrichum* Hedw.

POTTIACEAE Schimp., Coroll. Bryol. Eur. 24. 1855 (1856], nom. cons. prop. Type: *Pottia* (Reichenb.) Fümr. [This familiar name is antedated by Hyophilaceae.]

PRIONODONTACEAE Broth., Nat. Pflanzenfam. I(3): 763. 1905. Type: *Prionodon* C. Müll.

PSEUDODITRICHACEAE Steere & Iwats., Bryologist 78: 490. 1975 [1976]. Type: *Pseudoditrichum* Steere & Iwats.

Pseudoleskeaceae Schimp., Syn. Musc. Eur. 491. 1860, "Pseudoleskeeae." Type: *Pseudoleskea* B.S.G. = LESKEACEAE.

Pterigynandraceae Schimp., Syn. Musc. Eur., ed. 2. CXIII, 618. 1876, nom. rejic. prop., "Pterigynandreae." Type: *Pterigynandrum* Hedw. = ENTO- DONTACEAE.

PTEROBRYACEAE Kindb., Sp. Eur. Northamer. Bry. 15. 1896. Type: *Pterobryon* Hornsch.

Pterogoniaceae Schimp., Syn. Musc. Eur. 507. 1860, "Pterogonieae." Type: *Pterogonium* Sw. = LEUCODONTACEAE.

PterygophyUaceae Braithw., Brit. Moss-Fl. 3: 193. 1903, nom. illeg. Type: *Pterygophyllum* Brid., nom. illeg. ≡ HOOKERIACEAE.

PterygophyUaceae Rodw., Tasmanian Bryoph. 1: 10. 1914, hom. illeg. Type: *Pterygophyllum* Broth., hom. illeg. = HOOKERIACEAE.

PTYCHOMITRIACEAE Schimp., Syn. Musc. Eur. 241. 1860, "Ptychomitrieae." Type: *Ptychomitrium* Fümr., nom. cons.

PTYCHOMNIACEAE Fleisch., Musci Fl. Buitenzorg 3: 656. 1908. Type: *Pty- chomnion* (Hook. f. & Wils.) Mitt.

Pylaisiaceae Schimp., Syn. Musc. Eur. 518. 1860, "Pylaisieae." Type: *Pylaisia* Schimp. = HYPNACEAE.

RACOPILACEAE Kindb., Bot. Centralbl. 76: 85. 1898. Type: *Racopilum* P. Beauv.

REGMATODONTACEAE Broth., Nat. Pflanzenfam. 1(3): 1125. 1908, "Rheg- matodontaceae." Type: *Regmatodon* Brid.

RHABDOWEISIACEAE Limpr. in Rabenh., Kryptogamen-Fl., ed. 2. 4: 271. 1886. Type: *Rhabdoweisia* B.S.G.

RHACHITHECIACEAE Robins., Bryologist 67: 448. 1964. Type: *Rhachithecium* Broth, ex Le Jol.

Rhacocarpaceae Kindb., Bot. Centralbl. 77: 55. 1899, "Racocarpaceae." Type: *Rhacocarpus* Lindb. = HEDWIGIACEAE.

RHIZOGONIACEAE Broth., Nat. Pflanzenfam. 1(3): 614. 1904. Type: *Rhizogonium* Brid.

RHYTIDIACEAE Broth., Nat. Pflanzenfam., ed. 2. 11:475. 1925. Type:*Rhytidium* (Sull.) Kindb.

Ripariaceae Schimp., Coroll. Bryol. Eur. 53. 1855 [1856], nom. inval. Type: *Cinclidotus* P. Beauv. = POTTIACEAE. [There is no generic name on which to base this family name.]

RUTENBERGIACEAE (Broth.) Fleisch., Musci Fl. Buitenzorg 3: 725. 1908. Neckeraceae subfam. Rutenbergioïdeae Broth., Nat. Pflanzenfam. I(3): 786. 1906, "Rutenbergieae." Type: *Rulenbergia* Geh. & Hampe in Besch.

SCHISTOSTEGACEAE Schimp., Coroll. Bryol. Eur. 54. 1855 [1856]. Type: *Schist- ostega* Mohr.

SELIGERIACEAE Schimp., Coroll. Bryol. Eur. 22. 1855 [1856]. Type: *Seligeria* B.S.G.

SEMATOPHYLLACEAE Broth., Nat. Pflanzenfam. 1(3): 1098. 1908, nom. cons. prop. Type: *Sematophyllum* Mitt. [See notes under Macrohymeniaceae.]

SORAPILLACEAE Fleisch., Musci Fl. Buitenzorg 3: 847. 1908. Type: *Sorapilla* Spruce ex Mitt.

SPHAGNACEAE Dum., Anal. Fam. Pl. 68. 1829, "Sphagnideae." Type: *Sphagnum* L.

SPIRIDENTACEAE Kindb., Bot. Centralbl. 76: 85. 1898. Type: *Spiridens* Nees.

SPLACHNACEAE Grev. & Arnott, Mem. Wernerian Soc. 5: 442. 1824, "Splachnoïdeae." Type: *Splachnum* Hedw.

Symphyodontaceae Fleisch., Musci Fl. Buitenzorg 4: 1110. 1923. Type: *Sym- phyodon* Mont. = DALTONIACEAE.

Syrrhopodontaceae Fleisch., Musci Fl. Buitenzorg 1: 193. 1904. Type: *Syrrhopo- don* Schwaegr. = CALYMPERACEAE.

Systegiaceae De Not., Epil. Briol. Ital.33. 1869, nom. illeg., "Systegiacei." Type: *Systegium* Schimp., nom. illeg. = POTTIACEAE.

Tayloriaceae Schimp., Syn. Musc. Eur. 295. 1860, "Taylorieae." Type: *Tayloria* Hook. = SPLACHNACEAE.

TETRAPHIDACEAE Schimp., Coroll. Bryol. Eur. 37. 1855 [1856], "Tetraphideae." Type: *Tetraphis* Hedw.

Thamniaceae Mönk, in Rabenh., Kryptogamen-Fl. 4(Ergänzungsb.): 647. 1927, nom. illeg. Type: *Thamnium* B.S.G., hom, illeg. = NECKERACEAE.

THELIACEAE (Broth.) Fleisch., Musci Fl. Buitenzorg 4: 1485. 1923.

Leskeaceae subfam. Thelioïdeae Broth., Nat. Pflanzenfam. 1(3): 981. 1907, "Thelieae." Type: *Thelia* Sull, in Sull. & Lesq.

THUIDIACEAE Schimp., Syn. Musc. Eur 493. 1860, "Thuidieae."Type:*Thuidium* B.S.G.

TIMMIACEAE Schimp., Coroll. Bryol. Eur. 87. 1855 [1856]. Type: *Timmia* Hedw., nom. cons.

Tortulaceae Braithw., Brit. Moss-Fl. 1: 181. 1884, nom. illeg. Based on: *Tortula* Hedw., nom. cons. ≡ EPHEMERACEAE. [The type of Ephemeraceae was included.]

TRACHYPODACEAE Fleisch., Hedwigia 45: 63. 1906 [1905]. Type: *Trachypus* Reinw. & Hornsch.

Trematodontaceae De Not., Epil. Briol. Ital. 31. 1869, "Trematodontacei."Type: *Trematodon* Michx. = DICRANACEAE.

Trichostomataceae Schimp., Syn. Musc. Eur. 141. 1860, "Trichostomeae."Type: *Trichostomum* Bruch, nom. cohs. = POTTIACEAE.

VIRIDIVELLERACEAE Stone, Jour. Bryol. 9: 21. 1976. Type: *Viridivellus* Stone.

Voitiaceae Schimp., Coroll. Bryol. Eur. 5. 1855 [1856]. Type: *Voitia* Hornsch. = SPLACHNACEAE.

WARDIACEAE Welch, Bryologist 46: 26. 1943. Type: *Wardia* Harv. & Hook.

Weberaceae Braithw., Brit. Moss-Fl. 1:291. 1887, nom. illeg. Type: *Webera* Ehrh. ex Lindb., hom, illeg. = BUXBAUMIACEAE.

Weissiaceae Schimp., Coroll. Bryol. Eur. 7. 1855 [1856], "Weisiaceae." Type: *Weissia* Hedw. = POTTIACEAE.

Zygodontaceae Schimp., Coroll. Bryol. Eur. 39. 1855 [1856]. Type: *Zygodon* Hook. & Tayl. = ORTHOTRICHACEAE.

References

Brothems, V. F. 1924-1925. Musci. In A. Engler & K. Prantl, "Die naturlichen Pflanzen-famuilien." Ed. 2. 10, 11. Leipzig.

Crosby, M. R. 1974. Tóward a revised classification of the Hookeriaceae (Musci). Jour. Hattori Bot. Lab. 38: 129-141.

Ireland, R. R. 1968. Pleuroziopsidaceae, a new family of mosses. Jour. Hattori Bot. Lab. 31: 59-64.

Magill, R. E. 1977. Towards the conservation of four family names in Musci Taxon 26: 597-598.

——— 1978. Amendment to Proposal 440: conservation of Entodontaceae. Taxon 27 (in press).

Mitten, W. 1868. A list of the Musci collected by the Rev. Thomas Powell in the Samoa or Navigator's Islands. Jour. Linn. Soc. Bot. 10: 166-195. pl. 5-6.

Muller, C. 1847. De muscis nonnullis novis vel minus cognitis exoticis (Beschluss). Bot. Zeit. 5: 825-830.

Robinson, H. 1971. A revised classification for the orders and families of mosses. Phytologia 21: 289-293.

Sayre, G., C. E. B. Bonner & W. L. Culberson. 1964. The authorities for the epithets of mosses, hepatics and lichens. Bryologist 67: 113-135.

Steere, W. C. & Z. Iwatsuki. 1974. *Pseudodilrichum mirabile* gen et sp nov. (Musci: Pseudoditrichaceae fam. nov.), a unique moss from Great Bear Lake, Northwest Territories. Canad. Jour. Bot. 52: 701-706.

Welch, W. H. The systematic position of the genera *Wardia, Hydropogon,* and *Hydropogonella.* Bryologist 46: 25-46.

Wijk, R van der, W. D. Margadant & P. A. Florschütz. 1959-1969. Index Muscorum. Regnum Veg. 17, 26, 33, 48, 65.